A MAN ON THE MOON

A N D R E W   C H A I K I N

THE ODYSSEY CONTINUES

A MAN ON THE MOON

ANDREW CHAIKIN

II
THE ODYSSEY CONTINUES

*Commemorating
the 30th Anniversary
of the first landing on the moon,
July 20, 1969*

BY ANDREW CHAIKIN AND THE EDITORS
OF TIME-LIFE BOOKS, ALEXANDRIA, VIRGINIA

THE AUTHOR

Born in 1956, Andrew Chaikin grew up in Great Neck, New York, with a fascination for the heavens and space exploration. At age 12, he made his first visit to Cape Canaveral, where he was lucky enough to meet several astronauts, including Jim Irwin *(below)*. While studying geology at Brown University, he participated in the Viking missions to Mars at the NASA/ Caltech Jet Propulsion Laboratory. After graduating in 1978, he became a researcher at the Smithsonian's Center for Earth and Planetary Studies at the National Air and Space Museum in Washington. In 1980 he joined the staff of *Sky & Telescope* magazine, where he was an editor until 1986. Chaikin is now a contributing editor for *Popular Science* and has authored numerous articles for *Air & Space/Smithsonian, Discover, Popular Science, World Book Encyclopedia,* and other publications. He is a commentator for National Public Radio's Morning Edition and served as a consultant on the HBO miniseries *From the Earth to the Moon.* When he is able to take time out from writing, Chaikin pursues songwriting and performing. He lives in Arlington, Massachusetts.

CONTENTS

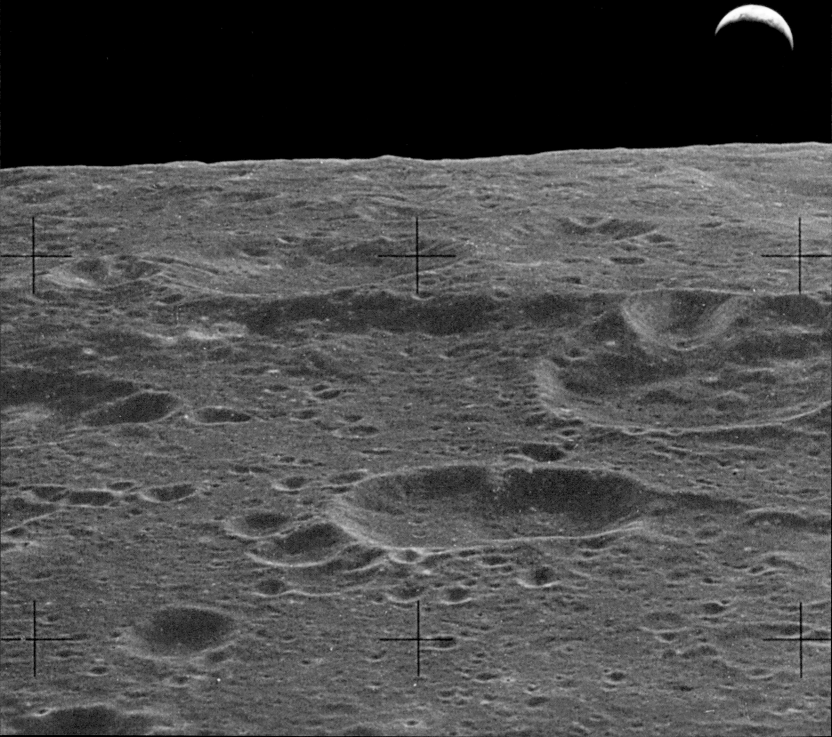

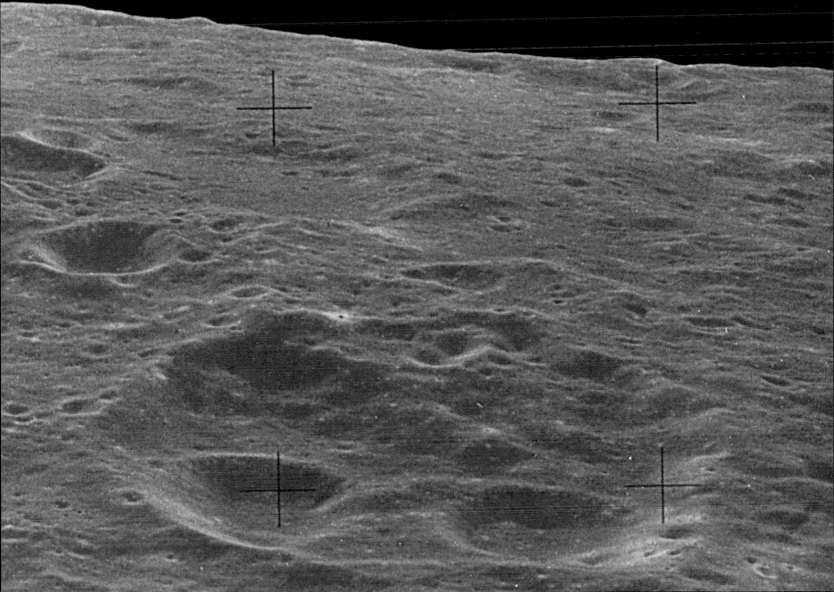

Bean stands near a color television camera set up to transmit scenes of the Apollo 12 moonwalks. Unfortunately, the camera ceased operating after Bean accidentally pointed it at the sun. His attempts to fix the damaged camera—including rapping it with his geology hammer—were of no avail.

Poised to collect lunar samples, Pete Conrad holds a pair of long-handled tongs for picking up small rocks. On their two moonwalks, Conrad and Bean gathered more than 75 pounds of lunar rocks, the largest about the size of a grapefruit.

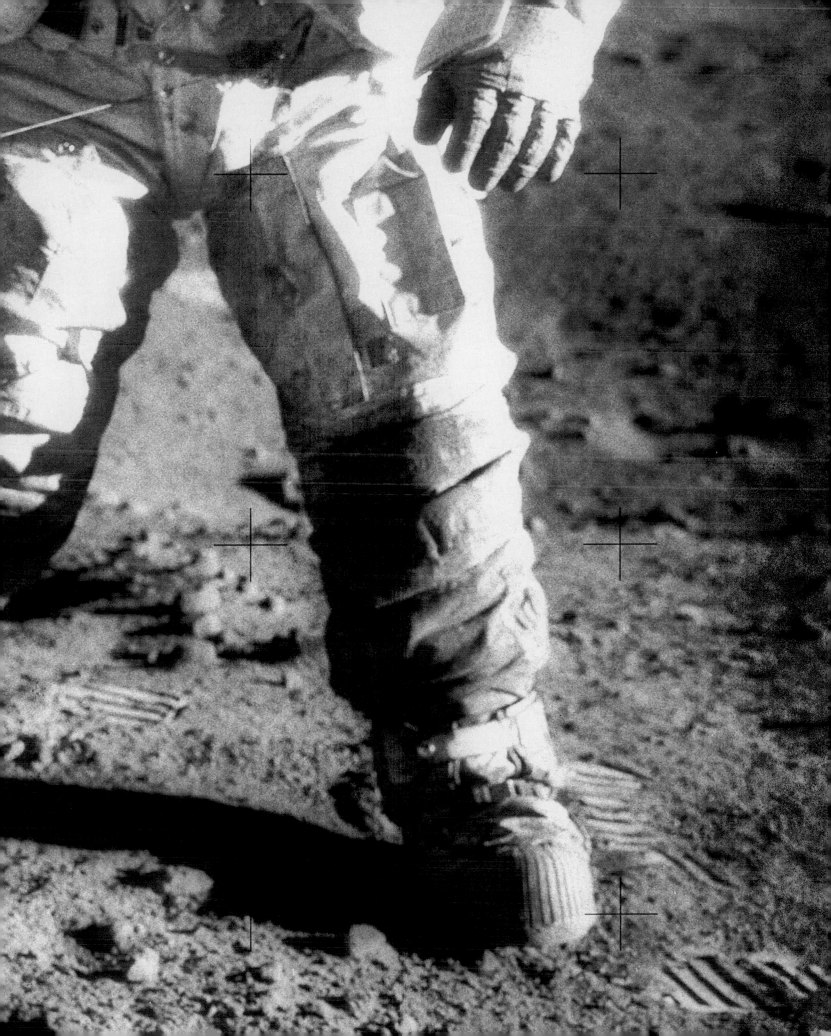

SAILORS ON THE OCEAN OF STORMS

APOLLO 12

I: THE EDUCATION OF ALAN BEAN

Al Bean, lunar module pilot for Apollo 12, holds a specimen container of soil collected from the moon's Ocean of Storms. His gold-plated sun visor reflects mission commander Pete Conrad as he snaps the picture.

When the men of Apollo 11 got out of quarantine, Richard Nixon had them and a few hundred guests out to Los Angeles to celebrate. By all accounts it was a hell of a party. Sometime well into the evening, one of the other astronauts in attendance—by now more than a little drunk—raised his glass. "Here's to the Apollo program," he said heartily. "It's all over." In a sense, he was right. John Kennedy's challenge to "land a man on the moon and return him safely to the earth"—the goal that had steered NASA for eight years—had been met. But was Apollo's mission over? Was the lunar landing simply an engineering demonstration, like the first flight across the Atlantic? There were those who thought so, even some within NASA. Kennedy had said nothing about a second lunar landing, or a third. You wouldn't ask Lindbergh to fly the Atlantic again, they said; why go back to the moon?

But in the summer of 1969 Apollo was only part of a much bigger question; the future of the space program as a whole was undecided. When Richard Nixon took office the agency had tried to gain early support for a manned space station in earth orbit. But the president deferred the matter

by creating a Space Task Group to formulate recommendations for NASA's future. By July, NASA administrator Tom Paine was working with the STG on a plan that picked up where John Kennedy had left off. Paine, like his NASA colleagues, believed that there was something implicit in Kennedy's challenge beyond its words, that it was a call for the United States to become a spacefaring nation. Apollo had given the country the technology to go to other worlds; it had only to exploit that capability. ☾

An expedition to Mars, envisioned here on the cover of Collier's magazine in 1954, was near the top of NASA's wish list at the time of the first Apollo landings. Tightening space budgets, however, ruled it out.

At NASA Headquarters, George Mueller and other planners had put together a far-reaching plan that Paine made even more ambitious in adapting it for the STG. The task group's timetable called for a twelve-man space station and a reusable space shuttle as early as 1975, depending on funding. By 1980 the station would have grown into a fifty-man space base; five years later there would be a hundred men in orbit. Meanwhile, there would be a base in lunar orbit by 1976, with a base on the lunar surface two years later. Then, as early as 1981, the first manned expedition to Mars would depart from earth orbit.

The plan was extraordinary—but it was not new. Almost twenty years earlier the same basic scenario had been mapped out in the pages of *Collier's* magazine by Wernher von Braun and other "space experts." At the time, von Braun was criticized for trying to sell the public a science-fiction vision of the future. In the summer of 1969, Paine was trying to turn von Braun's vision into reality. Like everyone at NASA, Paine hoped that the spectacular success of Apollo 11 would create a groundswell of support in the White House and in Congress to propel the space program onward and upward. Fortunately, Vice President Agnew, the chairman of the STG, was extremely enthusiastic, especially about the missions to Mars. The group's report would be ready for presentation to the White House in September. But already, there were signs it would not be well received. Since 1965, the year Apollo funding reached its peak, NASA's budget had steadily declined. Nixon staffers had told the agency that this trend would continue, and had indicated that even the first building block in the STG plan, the earth-orbit space station, would be a tough sell.

Meanwhile, Apollo moved on. Jim Webb had needed all his persuasive abilities to convince Congress and the Bureau of the Budget to pay for

☾ Throughout this volume, a crescent at the end of a paragraph signals an author's note at the end of the book.

enough Saturn Vs to fly missions through Apollo 20. He had done so on the premise that no one knew how many flights would be necessary to meet Kennedy's challenge. Now that the first lunar landing had been accomplished sooner than anyone expected, there was enough hardware to fly nine more lunar landings. Paine intended to make good on Webb's foresight. The handful of doubters aside, NASA had no intention of abandoning the moon. And if anyone wondered what would come from going back to the moon, they had only to be inside the windowless Lunar Receiving Laboratory at the Manned Spacecraft Center, on the evening of July 25, 1969.

In the LRL, five geologists dressed in white, hospital-style clothes and caps stood around a vacuum chamber. A big, powerfully built technician reached into a pair of space-suit arms attached to one side of the chamber to open a silvery container about the size and shape of a large tackle box. Inside, preserved in a lunar vacuum, were pieces of the moon. For the better part of a decade, geologists had labored to extend the disciplines of their science to an alien world by remote observation. What they had managed to learn about the moon from their telescopes and then, the unmanned probes, testified to the power of human intelligence. But they had always longed for the moment when they could probe the moon with their state-of-the-art laboratory instruments—or simply with their own eyes. ☾

After long moments of effort, the box was open. The technician removed a mesh covering, and a strip of foil that was part of a scientific experiment, and set them aside. Then he stepped away from the chamber and let the scientists look. With television carrying the event live, Harvard geologist Clifford Frondell peered into the chamber and blurted, "Holy shit! It looks like a bunch of burnt potatoes!" The rocks were so covered with charcoal-colored dust that the geologists couldn't tell anything about them. At this moment the curiosity that

Architect of the Saturn V moon rocket, Wernher von Braun also designed the Redstone missile *(below)*, a modified version of which launched the first U.S. satellite in 1958. Working to advance his vision of space exploration, von Braun helped create futuristic articles for *Collier's*.

Inside NASA's Lunar Receiving Laboratory in Houston, a scientist studies a piece of the moon collected by Neil Armstrong. This chunk, weighing about 1¼ pounds, is an agglomeration of rock fragments called a breccia.

gripped them was not scientific, but human: these were pieces of the moon.

Two nights later the first sample to be cleaned was raised up inside the chamber while Frondell and the other scientists watched. Instantly they recognized it as a piece of basalt. Familiar minerals glinted under the chamber lights. In the days and months to come, the scientists would try to coax secrets of lunar history from the rocks and dust of the Sea of Tranquillity. And this was just the beginning. The moon had been transformed from a light in the sky to a world ripe for exploration, and the geologists had already picked out candidate landing sites for the landings to come, an ambitious and spectacular roster of missions. One team of astronauts would set down at the edge of a huge, winding canyon called Schröter's Valley, where perhaps a billion years of lunar history might be exposed, layer-cake-style, in its walls. Another would visit Marius Hills, a field of low domelike bumps which the geologists hoped might be small, ancient volcanoes. The grand finale, for Apollo 20, might be a descent into the yawning amphitheater of Copernicus crater. These missions would be scientific feasts, with three-day stays on the surface and lunar roving vehicles or perhaps one-man lunar flyers. Their goal, as ambitious as any in science, was to answer the most basic questions about earth's nearest neighbor: How did the moon evolve? Was it really the cold, geologically dead world it seemed to be? Where did the moon come from? Solving these mysteries could open doors to even grander ones, for on that pockmarked, lifeless mass might be preserved the earliest history of the solar system, long erased on earth. The scientists' fondest hope was that the moon would tell human beings how their own planet came to be.

By November 1969, three astronauts were ready to open the door to these explorations with the flight of Apollo 12. For one of them, the moon represented the end of a long, private journey of waiting and perseverance.

FRIDAY, NOVEMBER 14, 1969

11:07 A.M., EASTERN TIME

PAD 39-A,

KENNEDY SPACE CENTER,

FLORIDA

With 15 minutes to go, reality grabbed hold of Alan Bean. Sealed in his space suit, lying on his back inside the command module *Yankee Clipper,* he felt his heart suddenly pound with anticipation. The spacecraft was on internal power now, and in the right-hand seat Bean had just put the fuel cells on the line; now he scanned the gauges for the electrical system. Over in the left-hand couch, Pete Conrad was talking to test conductor Skip Chauvin, nearing the end of the pre-launch checklist. Dick Gordon, in the middle couch, was making last-minute switch settings. It was very quiet; everything

Having just freshened his coffee, Pete Conrad talks with his crew over a pre-flight breakfast on the morning of November 14, 1969. Seated in the corner is Irving, a toy gorilla from Conrad's friend Peter Firestone of the tire-and-rubber family.

was just the way it had always been in the simulator. But this was no simulation. Silently, Bean told himself they were really going.

Already today, the crew of the second lunar landing had faced the threat of postponement because of the weather. Conrad, Gordon, and Bean had arrived at the pad under overcast skies. While they ran through the checklist a hard November rain lashed at the spacecraft atop its Saturn V booster. Rivulets of water found their way underneath the boost protective cover and danced across the command module windows. But the weather was erratic; the skies would seem to clear for a time and then gloom over again. Three and a half miles away, in the Launch Control Center, launch director Walter Kapryan deliberated and occasionally polled Houston for an opinion. Finally a report from an air force weather plane tipped the scales: ceilings acceptable, winds within limits, no lightning for 19 miles. They would go.

It was strange how unreal it had all seemed, up to now. For eight months Bean had been training to go to the moon and talking about it, and yet it was hard for him to believe it was really going to happen. Even this morning, as he sat down to a launch-day breakfast of steak and eggs with Conrad and Gordon, suited up, then rode out to the pad, it all felt like a practice run. But now, as cryogenic propellants flooded into the Saturn's tanks, Bean knew that the moment of fire and noise was almost upon him, and with it, the end of six long years of waiting. He had spent more time as a rookie than any astronaut in his group. Just why, Bean could only guess, and only in retrospect. But none of it mattered now.

Five minutes to go. "Pete, you guys have a good trip," radioed Skip Chauvin.

"Yes sir," said Conrad calmly, confidently. "Sure appreciate everything."

"Hold off the weather for five more, will you?" added Dick Gordon.

For Bean, every day of the past eight months had been an adventure, and the best part of it was training with Conrad and Gordon. The three of them had a bond that went deeper than their mission. They shared a history that began at naval air stations in Florida and California, and the test pilot school at Patuxent River, Maryland. A seat on a lunar mission was as much as any rookie could ask for, but going with Conrad and Gordon was almost too good to be true.

"One minute," Chauvin radioed. Pete Conrad put out his gloved hand and for a moment the three men clasped hands.

Chauvin called, "Thirty seconds, Twelve."

"Roger," said Conrad.

At 14 seconds, Chauvin began to count down. Seconds later the three men heard the distant rumbling of the Saturn's engines igniting. As Chauvin kept counting down, the spacecraft was engulfed by vibration. There was some noise, but not too much—they could still hear Chauvin's voice, counting down to zero, even as the vibrations increased. Somewhere in the midst of that commotion, Apollo 12 left the earth.

"Liftoff," called Conrad, "the clock is running." By now the whole spacecraft was shaking, but not as bad, Conrad thought, as he had expected. He scanned the instruments as Dick Gordon called out the time—"Three seconds . . . six seconds . . ." It seemed to take forever for the Saturn to rise past the launch tower, and when it did, Conrad radioed cheerfully to Houston, "That's a lovely liftoff!"

In the center seat, Dick Gordon glanced through a tiny window in the boost protective cover. "Everything's looking great," he told Conrad. "Sky's gettin' lighter."

Conrad kept a hawkeye on the 8-ball, the artificial-horizon indicator that showed the spacecraft's orientation. It registered the Saturn's slow roll as the beast steered onto the proper heading. Everything was right on schedule. It was like a perfect simulation. "Roll's complete," he announced.

Suddenly, out of the corner of his eye, Conrad was aware of a bright flash of light outside. At the same instant, a long burst of static filled his ears. He felt the spacecraft tremble. He spoke rapidly over the intercom, his voice flat: "What the hell was that?"

Only Conrad had seen the flash, but suddenly all three men heard the sound of the Master Alarm ringing in their headsets. Conrad glanced over to the center panel and was startled to see almost every light that had anything to do with the electrical system glowing brightly. He could hardly believe his eyes. Simulation upon simulation, and he'd never seen this many warning lights at once. Over the intercom, he started to read them aloud:

"AC BUS 1 light, all the fuel cells—I just lost the platform!" Conrad saw the 8-ball tumble aimlessly. More warning lights came on, confirming that the command module's navigation platform was out of commission. Conrad called Houston. "Okay, we just lost the platform, gang. I don't know what happened here; we had everything in the world drop out." Only a slight strain could be heard in his voice.

In the right seat, Bean was mystified. He had seen so many electrical crises in the simulator that he could, just by looking at the pattern of warning lights, recognize any given malfunction almost immediately. But he'd never seen so many lights before. The thought flashed through his mind that something had severed the electrical connection between the command module and service module. Had the emergency detection system sensed some problem with the booster, triggering the escape tower and whisking them away? No, he'd have felt a tremendous jolt. And yet, something was seriously wrong with the electrical system. Or was it? The meters showed that the spacecraft was still drawing electrical power, although at a lower voltage than normal. Could he even trust the gauges? "There's nothing I can tell is wrong, Pete."

Conrad keyed his mike and read the list of warning lights to Houston. He hoped mission control would be able to tell more than he could.

In mission control, a young flight director named Gerry Griffin heard Pete Conrad describe, in one breath, the longest list of malfunctions he had ever heard. Griffin couldn't believe this was happening, not on his first mission as a flight director. He was certain he'd have to abort the flight. His voice calm, Griffin called on John Aaron, the bright, twenty-four-year-old flight controller in charge of the electrical system, and heard only silence. Griffin called again: "What do you see?" Aaron's problem was that aside from a bewildering maze of warning lights on his console, there wasn't anything *to* see. On his screen, telemetry from the spacecraft had been replaced by an undecipherable pattern of numbers. But Aaron had encountered the same problem during a practice run the year before, and he knew how to fix it. He could visualize the command module instrument panel, and the switch, labeled Signal Condition Equipment, that the astronauts had to set to recover the data. He said quickly and confidently, "Flight, try S-C-E to Aux." ❨

This control was so obscure that Gerry Griffin had no idea what it was; neither did Capcom Jerry Carr as he radioed the request to Apollo 12. And neither did Pete Conrad. It was Al Bean who knew where to find the switch, and moments later, Aaron had his telemetry back. For some unknown reason, the spacecraft's fuel cells had been knocked off-line. Unless they could be reconnected, the command module would have only its batteries for power, and they were reserved for reentry. At the flight director's console, Gerry Griffin weighed the possibility of an abort.

Still the Saturn sped onward. Conrad glanced up and saw bright sunlight. They were above the clouds; they were heading in the right direction. Whatever had happened to them, it hadn't touched the booster or its guidance system. They'd make it into orbit, but Conrad was afraid he'd wind up with a dead spacecraft when they got there.

The g-meter had passed 3 and was still climbing. Now the intercom buzzed with excited voices; thick with the weight of acceleration: "Try the buses—get the buses back on the line. . . . I've lost the event timer. . . . Two minutes, EDS AUTO. . . ."

Bean heard Carr say, "Apollo 12, try and reset your fuel cells now." Bean was reluctant to do that without knowing what had gone wrong. Conrad said, "Wait for staging." They were coming up on that chaotic moment when the first stage would drop off and the second-stage engines would ignite. Conrad and Gordon told Bean, "Hang on!"

The first stage fell away and the three men were slammed against their harnesses, just as they had expected. Seconds later, Bean revived the stunned fuel cells. Everything was back on line, apparently none the worse for wear. As the second stage did its work, Conrad offered a theory: "I'm not sure we didn't get hit by lightning."

Conrad was right. As it lifted off, the Saturn had trailed a column of flame and ionized gases which stretched all the way to the ground. Tearing through the rain clouds it became the world's longest lightning rod. Thirty-six seconds after liftoff, a bolt of electricity discharged right through Apollo

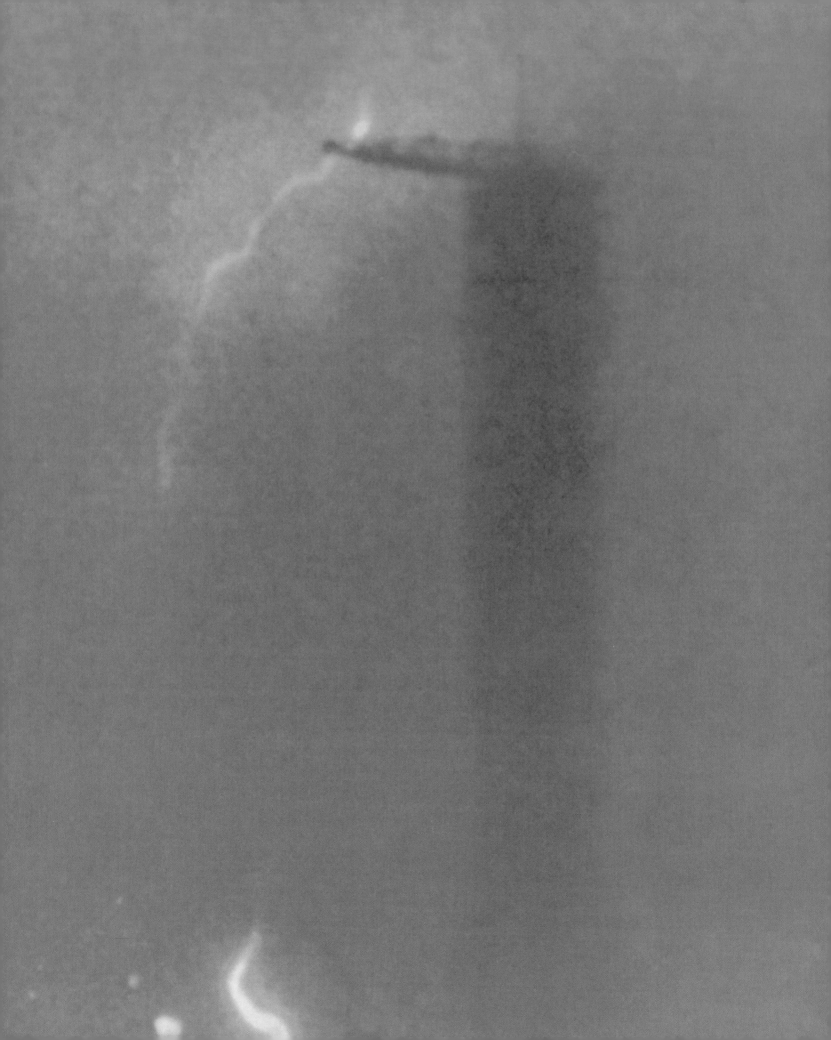

12 and onto the launch tower 6,000 feet below. The command module had shut itself off in response to the tremendous electrical surge. A second strike at 52 seconds, unnoticed by Conrad and his crew, had wiped out the navigation platform. Both lightning bolts were recorded by an automatic movie camera near the launch pad, as analysts would later discover.

Now, as Bean reconnected the fuel cells, the warning lights blinked out one by one. The platform was still out, but they would deal with that once they were in orbit. "Twelve, Houston," called Carr. "You're right smack-dab on the trajectory." The second stage continued its long, smooth push, and inside *Yankee Clipper* tension evaporated. Pete Conrad let out a giddy, high-pitched giggle, like a schoolboy who had just sneaked out of class without getting caught. Then he laughed, and Gordon and Bean laughed with him—

"Was that ever a sim they gave us!"

"There were so many lights up there I couldn't read them all!"

—and they laughed all the way into orbit. ☾

●◑◐○○○◐◐

To Pete Conrad, the launch of Apollo 12 exemplified one of the most important differences between flying airplanes and flying in space. Conrad could remember narrowly avoiding a midair collision with another plane when he was at Pax River; after the miss his heart pounded for several minutes. But during the lightning strike it was different. There was no terror, nor would

Jane Conrad, with son Christopher, watches her husband's spacecraft disappear into rain clouds *(below)*. Seconds later a bolt of lightning struck the rocket, setting off a flurry of alarms in the command module, then streaked along the rocket's tail of exhaust gases to zap the launch tower *(left)*.

there be at any other time in Conrad's four spaceflights, because things just didn't happen fast enough for that. And it proved one of spaceflight's emerging maxims: If you don't know what to do, don't do anything. Conrad never gave the abort handle a moment's thought.

But now, in orbit, Conrad worried that the lightning had damaged something, and that Houston was going to call off the mission. They'd ride around the earth a couple of times, and then they'd have to go home. The first order of business was to check out *Yankee Clipper*'s systems, and as soon as Apollo 12 reached orbit Conrad put his crew to work. Dick Gordon went down in the lower equipment bay, sweating through an effort to realign *Yankee Clipper*'s navigation platform with the stars. He kept talking to Bean, saying, "I don't see anything, Al," and for a while he wondered—what happened to the *stars*? Finally he realized he hadn't given his eyes a chance to adapt to darkness. Bean, consulting the star chart, told him he ought to be seeing the constellation Orion, and sure enough, there it was in the telescope, with brilliant Sirius nearby. Gordon made the alignment, with little time to spare.

During breaks in the work, Conrad and Gordon gave Bean a guided tour of earth orbit. "Here comes sunrise, Al," they said, or, "Hey, there's an island down there," and then in darkness over Africa, "Look down there; those are campfires," and he looked up from his checklist to witness the most amazing sights of his life. But for Bean, as well as his veteran crewmates, the lightning strike had been just as memorable. ☾

"That'll give them something to write about tonight," said Conrad, thinking of the press. "I'll bet your wife, my wife, and Al's wife fainted dead away."

Gordon said, "I bet they did when you started calling out about eighteen lights."

"Every time I close my eyes," Conrad said, "all I see are those lights." For Conrad, Gordon, and Bean the launch of Apollo 12 had already become a sea story that they would tell again and again, for the rest of their lives.

To their relief, everything on the command module (and the lander, which was undergoing scrutiny in Houston via telemetry) seemed to be checking out perfectly. At 2 hours, 28 minutes Jerry Carr radioed the message they'd been waiting for: "Apollo 12, the good word is you're Go for TLI."

"Whoop-de-do!" crowed Conrad. "We're ready! We didn't expect anything else."

What mission control did not tell Conrad was that they feared the lightning had damaged the pyrotechnic system used to deploy the command module's parachutes. Chris Kraft in Houston had conferred with other NASA managers at the Cape, and they decided to continue the mission. The rationale was simple: Conrad and his crew would be just as dead if the parachutes didn't work now as they would after coming back from the moon, 10 days from now.

Minutes later Conrad, Gordon, and Bean were strapped in, waiting for the third stage to relight. When it did, all three men felt it rumbling away, speeding them out of earth orbit. Conrad said, "Al Bean, you're on your way to the moon."

"Yeah," Bean replied. "Y'all can come along if you like."

●◑○○○○◐●

There was a picture on the wall of Dick Gordon's study in Nassau Bay, a signed photograph of a young, flight-suited Pete Conrad—with a bit more hair—posing next to a Phantom jet. Conrad was about to leave Miramar Naval Air Station in San Diego to report to his new job as an astronaut. By the time that photograph was taken Conrad and Gordon had become best friends. They had shared the snap of the catapult on the deck of the carrier *Ranger*, where they had roomed together and flown missions with their squadron. Their friendship was cemented in weeks at sea and in the nightspots of San Diego. And in the summer of 1962, they had shared the hope of making it into the second astronaut selection. Gordon didn't find out that Pete had succeeded until the public announcement in the middle of September. Soon after, Gordon got a letter from Robert Gilruth, the director of the Manned Spacecraft Center, saying that he was sorry, but although his qualifications had been excellent . . .

Gordon's feelings could not have been more divided: joyful that Pete had made it; heartbroken that he wasn't going with him. Maybe Pete was just being his optimistic self, or maybe he had a premonition of how things would someday turn out. But when he gave Gordon that portrait, he wrote on it, "To Dick: Until we serve together again."

It was no exaggeration to say that the high point of Dick Gordon's life were the three days he and Pete Conrad spent in orbit on Gemini 11. Even before launch, they had earned a reputation as the cockiest, not to mention the most fun-loving, team of astronauts ever to fly: Conrad, short, balding, and wisecracking, and Gordon, not much taller, but more formidable, with the rugged face and build of a boxer. Gordon's friends knew him as a man of strong emotions. He was such a ladies' man that Conrad—much to Gordon's dislike—took to calling him "Animal." ☾

Unlike many astronauts, Dick Gordon had not had boyhood dreams of being a pilot. Growing up in Seattle, his family hard hit by the Great Depression, he had dreamed of the priesthood. In college he majored in chemistry but considered pursuing a professional baseball career. Finally he settled on dentistry. But soon after graduation his life was interrupted by the Korean War, and it was only then, in the navy, that Gordon found his calling. Gravitating toward the excitement of aviation, he was hooked after his first training flight. After the war, flying with a squadron that had the reputation of being the cockiest in the navy, Gordon won top honors for his precision in the maneuvers used to deliver nuclear weapons. Then it was on to the test pilot school at Pax River, where his friendship with Pete Conrad began. Gemini 11 solidified the bond between the two men, on duty and off. They thought alike; they flew alike. In the simulator or in a T-38, they anticipated each other's moves as if they were communicating by telepathy.

Everything about them said, "I'm the best—see if you can get one up on me."

By the time Conrad and Gordon lifted off in November 1966, they already had enough stories to last a long time—and there were more waiting for them in earth orbit. An hour and a half after launch Conrad and Gordon shared the triumph of the first one-orbit rendezvous and docked with an Agena target rocket. A day later, Gordon climbed outside Gemini 11, trailing a thirty-foot umbilical and faced the most dangerous experience of his life. His assignment was to attach a Dacron tether from the Agena to the nose of the Gemini in preparation for a later experiment. But his body kept floating off the Gemini, and with no means of anchoring himself he worked so hard that he overloaded the cooling system in his suit. By the time he had attached the tether his heart rate had reached 180 beats per minute. His breathing was heavy; his eyes stung with perspiration. Inside Gemini 11, Conrad was gripped with concern for his partner. He knew that if Gordon became incapacitated there would be no way to pull the man, in his pressurized suit, back into the tiny cabin; Conrad would have no choice but to cut him loose and go back without him. Unwilling to order his copilot back

A space-walking Dick Gordon struggles to attach a tether from the nose of Gemini 11 to an Agena target rocket. Gordon became so exhausted during the task that Conrad feared he might lack the strength to reenter Gemini's cabin.

Pete Conrad shouts to Dick Gordon, his Gemini 11 crewmate and best friend, in an effort to cut through the din on the flight deck of the recovery carrier U.S.S. *Guam* in 1966. "To Pete:" Gordon inscribed the picture, "After three days I thought you'd stop yelling at me. P.S.—I'm still listening."

inside, he waited, hoping Gordon would make that decision on his own. Conrad would remember those moments as the scariest he had ever experienced in a spaceflight, and that included the lightning strike on Apollo 12. But the flight also had its funny moments. On the third day Gordon took a second, much more leisurely space walk; this time he merely stood up in the hatch and shot astronomical photographs. During a lull, as Gemini 11 drifted in daylight over the Atlantic, Conrad fell asleep, his arms sticking out in front of him in his pressurized space suit. He awoke with a start and said to Gordon, "Hey, Dick, would you believe I fell asleep?" To which Gordon answered, "Huh? What?" He'd been asleep too, standing in the void, whizzing around the world at 17,000 miles an hour.

When Pete Conrad moved on to Apollo, there was no question that Gordon would come with him. Under other circumstances, the two men might have landed on the moon together, but Deke Slayton had a rule that command module pilots, like commanders, had to be veterans of space rendezvous. But the lunar module pilot's seat could be filled by a rookie, and Pete Conrad knew just who that rookie shoud be.

● ● ◐ ○ ○ ○ ○ ◐ ●

From the very beginning, Alan Bean saw the world differently from other test pilots. Long before he ever flew, as a boy in Fort Worth, Texas, he was captivated by the beauty of the airplane in motion. In the summer of 1956, the twenty-four-year-old Bean, thin and bright-eyed, with the well-toned body of a college gymnast, arrived at Jacksonville Naval Air Station. He was the youngest and newest member of attack squadron VA-44. In a group photo taken sometime afterward, two things about Bean stand out: his ear-to-ear grin and his tie, which is almost the only one that's perfectly straight. Even as a squadron pilot, Bean's sense of aesthetics was very much intact. He longed to fly the best looking planes, like the sleek F-8U fighter. Unlike the other pilots, who spent their free time fixing up old cars, Bean took night classes in oil painting.

Bean had always been a loner, but he felt close to the fliers in VA-44. It was a place where everyone had a nickname: there was Dancing Bear and Rockie Pie, Spanky, Dinky, Tiger Jack, and Von Du Quick. Sometimes they called Bean "Sarsaparilla" because he never touched a drink; sometimes they just called him "Beano." Once, on liberty in Paris, the pilots went drinking and decided to put Sarsaparilla to good use. While they drank inside, Bean did gymnastics outside. Before long he drew a crowd. His squadron mates then came out, one by one, to look for a date.

As a gymnast at the University of Texas, Alan Bean learned the value of hardworking perseverance and a positive attitude—useful lessons for the six years he would serve as an astronaut before a moon mission came his way.

When it came to flying, Bean wasn't a "natural" like some of his squadron mates who seemed born to be great aviators. But whatever Bean lacked in innate talent he made up in effort. Through sheer determination he mastered the subtleties of precision bombing from a speeding jet and turned himself into one of the best weapons delivery pilots in the squadron. After a two-year tour at Pax River, Bean felt he'd reached the pinnacle of his profession. And while he shared the other fliers' drive to be the best, he was always aware that his outlook differed from theirs, even after he left Pax River for another attack squadron in Cecil Field, Florida. In 1962, during the Cuban Missile Crisis, there was talk of sending Bean and his colleagues into combat. The other pilots wanted it so bad they could taste it—after all, this was what they had been trained for. Bean would have gone in a minute, to defend the interests of his country, but he was hardly excited about it. There was no love of combat in his soul, only a love of flying.

By that time, Bean had his sights on a new goal: to become an astronaut. At the suggestion of one of his former instructors from Pax River, he applied for the second selection in 1962. He made it into the final group of thirty-five pilots, but no further; the instructor, whose name was Pete Conrad, went all the way. The following year Bean applied again; this time he made it.

At NASA, Bean was surprised at how different it felt to be in the Astronaut Office than it had in the squadron or at Pax River. Camaraderie was over-shadowed by competition for flights. And for some of the rookies, Al Shepard was an ominous presence. To Bean, Shepard seemed like a tiger shark swimming around in a tank full of fish. He didn't go after the big fish; he liked to take a bite out of a minnow every now and then. For quite a while Bean felt like a minnow. And once again Bean felt he saw things differently from the other pilots. The Old Heads' animosity for the doctors, for example, made no sense to him. So what if they didn't like Chuck Berry; did that mean they shouldn't cooperate with him? Another day it might be some PR requirement—the television people would want to film astronauts in train-ing, and the Old Heads would grumble. Bean thought, "What's the problem? We won't even know they're there." And he didn't hesitate to say so in front of Shepard and Slayton. Nor did Bean share the Old Heads' view of the non-test pilots in the office. He gravitated to Anders, Cunningham, and Schweick-art precisely because they *were* different. They thought about things other than flying and work. They examined their lives. Rusty Schweickart taught him to understand points of view very different than the ones he had learned in the military. He marveled at Bill Anders's intellect, not only about science but about the ways of the world. And he admired outspoken Walt Cunningham for his directness. For Bean, being with the three of them was an education. ❰

But Bean wasn't learning the right lessons about being an astronaut. He followed Bill Anders's motto (Work hard and someone will notice), and when he was named as backup commander on one of the last Gemini mis-sions, he thought his day had finally come. Never mind that Gemini would be over before he had a chance to fly; he was sure Slayton and Shepard had recognized what he could do and that they were grooming him for Apollo. But in the fall of 1966, he got his assignment: astronaut representative on the Apollo Applications Project, the space station planned for earth orbit in the 1970s. Bean's heart sank as he saw his chances for a lunar mission evaporate. While he was flying a desk, Anders, Cunningham, and Schweickart were get-ting ready to fly in space, and maybe even to walk on the moon. ❰

●●◖○○○◗●●

In the fall of 1966, when Deke Slayton asked Pete Conrad to pick a lunar module pilot, he requested his old student from Pax River, Alan Bean. Slayton told Conrad that Bean was unavailable; he was assigned to Apollo Applications. Why Bean was put on AAP, Conrad had no idea, but in his own mind it was a bad deal. He couldn't help but wonder whether somebody had put Bean there to get him out of the way. It wouldn't have been the first time Bean had gotten in trouble, and Conrad had been around for that too. In

A pensive C. C. Williams peers from the Gemini simulator while training as backup pilot for Gemini 10 in 1966. The following year, Williams' life ended tragically in the crash of his T-38 jet, the result of a mechanical failure.

1960, Bean arrived at Pax River and became one of Conrad's students. Conrad had met the younger man four years earlier in Jacksonville, where they flew in sister squadrons. At Pax, Bean was quieter and more serious than many of the other pilots Conrad taught. He wasn't the best aviator in his class, but he was near the top, and definitely one of the brighter students. The thing Conrad noticed most about Bean was his persistence. Once he latched onto a problem he was absolutely tenacious about finding a solution, and he wouldn't let go until somebody listened. Bean also tended to speak his mind, a little too much for his superiors. When it came time to give the graduates their assignments, Conrad found out that the brass were planning to send Bean to Electronics Test, the boondocks of test flight. Conrad went to the school commander and argued Bean's case for better duty, and his intervention won Bean a place in Service Test, where pilots wrung out new airplanes before clearing them for squadron duty. For a pilot of Bean's caliber it was a far better place to be. ◖

But in 1966 there was nothing Conrad could do to rescue his former student. Slayton wasn't letting him have Bean, and that was that. So Conrad chose another astronaut he knew, a big, friendly bear of a marine named C. C. Williams, who had also been one of his students at Pax. Williams stood just over six feet tall. During the third astronaut selection he made it under the height limit—just barely—by spending the night before the physical jumping up and down to compress his spine. Williams brought a gentle, self-effacing presence to the astronaut corps. He would set his jaw and utter with mock seriousness, "I'm a marine. I'm a trained killer. . . ." As the first bachelor in the Astronaut Office he was the envy of his colleagues. This distinction ended when he married a former water-skiing performer from Florida's Cypress Gardens. Williams fit in well with Conrad and Gordon, and the three of them leaped into training as the backup crew for the first manned lunar module flight.

●◗◖○○○◗●

Meanwhile, Alan Bean toiled in the backwaters of Apollo Applications, oblivious to Conrad's efforts. In the navy, he had learned to meet adversity with a positive attitude. His commander in Service Test at Pax River would always remember Bean for taking an unpleasant job and doing it better than any junior officer he'd ever seen. And it turned out that when NASA put out a call for new astronauts, the man was on the navy's astronaut selection committee. He made sure Bean was on the list. Bean made up his mind that he would do the best job on AAP that anyone had ever seen.

But as time passed, Bean began to see that doing a great job wasn't enough if the right people didn't know about it. Slayton and Shepard wouldn't notice him unless he learned to promote himself. And he'd have to make it look unintentional; that would take some work. Some of the Fourteen were masters at it; Bean would study their technique. And he would learn when to keep his mouth shut. Playing politics was abhorrent to him, but he had no choice.

But these realizations had come a few years too late. No one knows how long he would have stayed there in what Pete Conrad called Tomorrowland, if fate had not intervened. On October 5, 1967, C. C. Williams was flying a T-38 from the Cape to Mobile, Alabama, to see his father, who was dying of cancer. Near Tallahassee the airplane suddenly went into an uncontrollable aileron roll. The jet lost all its lift and became a ballistic projectile, accelerating earthward at a horrendous rate. Williams followed the procedures for emergency ejection, but the T-38 was going so fast that he was already far too low for his parachute to save him.

Not long afterward, Bean was at Ellington and ran into Pete Conrad on the flight line. Since coming to NASA, Bean hadn't really had much contact with Conrad. They were in different astronaut groups, and aside from the pilots' meetings, there had been only occasional, brief encounters. Now his old teacher said, "Have you got a minute, Al?" Conrad had told Slayton he wanted Bean as his new lunar module pilot; Slayton had agreed. Bean couldn't believe what he was hearing. Deliverance had come.

● ◐ ○ ○ ○ ○ ◑ ●

They called him Beano, and he fit right in. His quiet seriousness complemented Conrad and Gordon's bravado. He didn't race cars, as they did. When it came to training, Conrad and Gordon had a certain restlessness about them—Let's get on with it, that was their style—but Bean delved into the writing of checklists as if it were his only profession. He sweated the details Conrad and Gordon hated. Even his eating habits were single-minded; he ate spaghetti almost every night of the week. They gave him all the razzing due him as the rookie on the crew—but they respected him, and each other.

They soon became a fixture at the Cape, where they were backing up Jim McDivitt's crew for the first LM mission. Now that he was on a crew, Bean realized that his whole outlook had changed. The every-man-for-himself undercurrent of the Astronaut Office no longer applied to the three of them. They were in this together, and they were going to fly the best mission ever flown. Together, they had more camaraderie than just about any Apollo crew.

That didn't mean the competition was over; to the contrary, it gave the men a chance to fuel the friendly inter-service rivalry that permeated the astronaut corps. They were the first all-navy Apollo crew, and Conrad was always glad for a chance to get some mileage out of it. McDivitt's crew was all air force, and Conrad would say things like, "Well, Jim, the navy's always glad to help out the air force . . ."

Then there was that fateful day in 1968 when, in the plan to send Apollo 8 to the moon, McDivitt's crew swapped places in line with Frank Borman's. Bean still remembered how Pete and Dick reacted. It was a bad deal, they said. Until then, it looked as if they might have a shot at the

---

Bean couldn't believe what he was hearing. Deliverance had come.

---

first lunar landing, but this was going to screw all that up. And Bean was just so happy to be on a flight—any flight—that it never occurred to him to think about being first or second to land on the moon.

By the fall of 1969, now training for the second lunar landing, Conrad, Gordon, and Bean had been at the Cape so long they seemed like permanent residents. Conrad had gotten to know a Cocoa Beach car dealer named Jim Rathmann, a former Indianapolis 500 winner who made fast friends with the astronauts from the Original 7 on down. Thanks to Rathmann, General Motors gave the astronauts great deals on sports cars, and sometime during the training for Apollo 9, Conrad had worked a deal for three matching gold Corvettes, and had Rathmann's shop customize them with futuristic black trim and a small "CDR" for mission commander Conrad, "CMP" for command module pilot Gordon, and "LMP" for lunar module pilot Bean. The security guards got used to seeing the three Corvettes tool out the gate of Patrick Air Force Base in the early evening and head down route A-1A for Cocoa Beach. They were like a bunch of squadron buddies. They shared each other's lives as well as work. In the morning, before a simulator run, the conversations sounded like the ready room at the squadron—"Let me tell you what happened last night!" You wouldn't find the backup crew, Dave Scott, Al Worden, and Jim Irwin, telling jokes about adventures from the

Having the time of their lives, moonward-bound Conrad *(left)*, Gordon, and Bean show off their customized Corvettes, acquired from a Cocoa Beach, Florida, Chevrolet dealer. Each car was personalized with its driver's crew position: CDR for commander, CMP for command module pilot, and LMP for lunar module pilot.

night before when they came in for work. And it was that way because Conrad set the tone. Conrad was the glue.

In the simulator, Conrad was spectacular. The instructors would throw everything in the book at him, and there wasn't anyone who could solve problems faster, or react quicker. If Conrad had one weakness, it was his language. When decorum was required, Conrad carried himself with all the poise of a Princeton-educated navy officer, but most of the time he raised colorful to an art form. In the simulator, he whistled and hummed and cracked his chewing gum so loudly that the instructors winced under their headsets. And when the malfunctions got thick, he swore like a sailor. The instructors smiled and shook their heads—"What's this guy gonna do during the *flight?*"—and in the next minute they'd break up laughing, because they'd hear Bean's quiet voice in the background: "Yep, that's astronaut talk. A-OK. I gotta learn that."

The instructors' worries were nothing compared to those of the NASA Public Affairs people. When it looked as if Conrad might have the first landing in his pocket, they chuckled at the thought of this little wisecracking, balding guy with the gap-toothed grin as the first man on the moon. Now, faced with the prospect of Pete Conrad on the airwaves, they were decidedly uneasy. They had barely recovered from the criticism that followed Gene Cernan's profanity during Apollo 10. (Never mind that Cernan had been in a life-or-death crisis a quarter of a million miles from earth; the letters came anyway.) And they weren't the only ones who were worried. One day during training Bean said to his commander, "You know, you can't talk like that on the flight," and Conrad replied that it was people like himself, who swore a lot on the ground, who never slip up over the air during a flight. Furthermore, he told Bean, people like you, who hardly ever swear, are the ones who do. That was the end of it, but Conrad had to laugh at the thought that Bean would worry about that until the end of the mission.

● ● ○ ○ ○ ○ ○ ◐ ●

Pete Conrad had been disappointed not to fly the first lunar landing, and there were plenty of NASA people, at the Cape and in Houston, who were surprised when he didn't get it. He was in mission control when Neil Armstrong and Buzz Aldrin landed on the moon. And he was there when everyone, from the NASA managers in the control center to the geologists in a back room to Mike Collins in lunar orbit, were trying to figure out *where* Armstrong and Aldrin had landed. Sam Phillips turned to Bill Tindall and said, "Next time, I want a pinpoint landing."

Phillips's request was understandable; there was no point in having the geologists painstakingly choose a landing site when there was no certainty of being able to get there. But Tindall thought it was impossible. After all, *Eagle* had landed four miles away from its aim point. The trajectory people had identified several sources for the error, and had figured out how to prevent them from recurring, but one unknown persisted, namely the moon's lumpy gravity field. Even now, no one knew exactly what mascons would do to the lunar module's path, so predicting its precise trajectory ahead of time was impossible. But when Tindall convened the trajectory experts, a young mathematician named Emil Schiesser made a breakthrough. The key was the Doppler effect, the apparent shift in frequency of light waves or sound waves emitted from a moving object as detected by a stationary observer. You can experience the Doppler effect standing next to a highway:

When decorum was required,
Conrad carried himself with all the poise of a
Princeton-educated navy officer, but most of the
time he raised colorful to an art form.

---

the horn of a passing car seems to rise in pitch as it speeds toward you, then fall as it moves away. The same phenomenon changes the apparent frequency of radio waves received from a spacecraft moving toward or away from the tracking stations on earth. Tiny Doppler shifts were already being used by controllers to analyze the trajectories of Apollo spacecraft during their lunar voyages. ❬

Radio signals from an LM in lunar orbit, Schiesser pointed out, have a predictable pattern of Doppler shift. The effect is strongest when the lander is flying over the edge of the moon as seen from earth, and weakest when it is over the geographic center of the near side. If planners could predict the pattern of Doppler shifts, they could compare that information with the actual shifts they detected. The differences would in turn reveal whether the lunar module was off course, and by how much.

Schiesser's idea was brilliantly elegant, but it left the problem of how to give that information to the astronauts in a form that would be easy to feed into the LM's guidance computer. Soon that answer too emerged in Tindall's meetings. The solution was to fool the computer into thinking the landing

STUCK IN THE STARTING BLOCKS

Less than a year after President Kennedy aimed
America's space effort at the moon, the Soviet
Union declared that it, too, would land a cosmonaut there—and do it first. Now the space race had
become a moon race. Given the Soviets' historic
firsts in space dating from Sputnik 1 in 1957, it
seemed in 1962 that they might deliver on their
promise. As it happened, however, the Soviet lunar
program hardly got off the ground. Much of the
fault lay with the huge N-1 moon booster *(right)*.
With 30 engines developing nearly 10 million
pounds of thrust, the rocket failed each of its four
flight tests. A few heartbeats after this picture was
taken, the booster began to roll uncontrollably,
then came apart in a huge fireball.

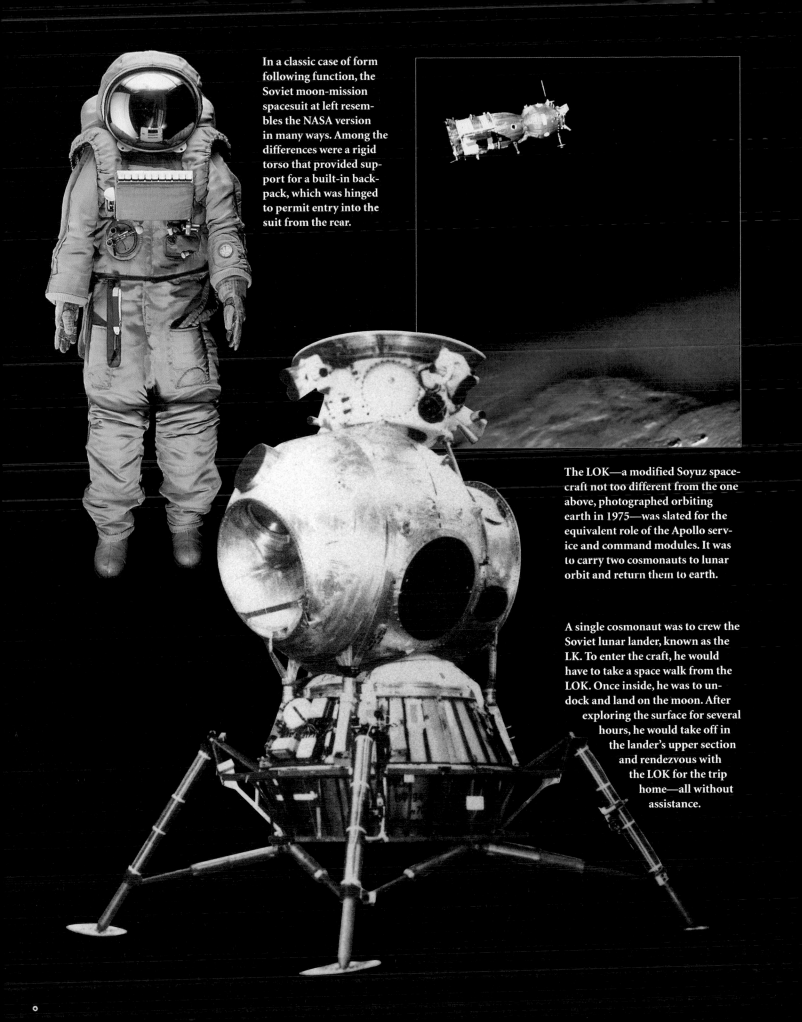

In a classic case of form following function, the Soviet moon-mission spacesuit at left resembles the NASA version in many ways. Among the differences were a rigid torso that provided support for a built-in backpack, which was hinged to permit entry into the suit from the rear.

The LOK—a modified Soyuz spacecraft not too different from the one above, photographed orbiting earth in 1975—was slated for the equivalent role of the Apollo service and command modules. It was to carry two cosmonauts to lunar orbit and return them to earth.

A single cosmonaut was to crew the Soviet lunar lander, known as the LK. To enter the craft, he would have to take a space walk from the LOK. Once inside, he was to undock and land on the moon. After exploring the surface for several hours, he would take off in the lander's upper section and rendezvous with the LOK for the trip home—all without assistance.

point itself had moved, rather than the lander. That change required entering only a single number.

At the same time, mission planners deliberated on where to send Apollo 12. They could simply have picked out a specific crater, but planning coordinator Jack Sevier had a better idea. The unmanned Surveyor 3 probe was perched on the vast lava plain called the Ocean of Storms, where it had landed in April 1967. The geologists had already identified the Ocean of Storms as one of their candidate targets for the second landing; they suspected its rocks would be younger than, and perhaps chemically different from, those in the Sea of Tranquillity. Now Surveyor 3 became the target for the first pinpoint lunar landing. It was a bold decision to commit the system to such a visible measure of success or failure; if Conrad and Bean missed, everyone would know it. But that was exactly the point: This goal would drive NASA to achieve what had first seemed impossible.

Conrad and Bean were to spend 31½ hours on the moon, some 10 hours longer than Armstrong and Aldrin. During that time they would take two moonwalks, each lasting about 3½ hours. Most of the first excursion would be devoted to setting out an autonomous scientific station called the Apollo Lunar Surface Experiment

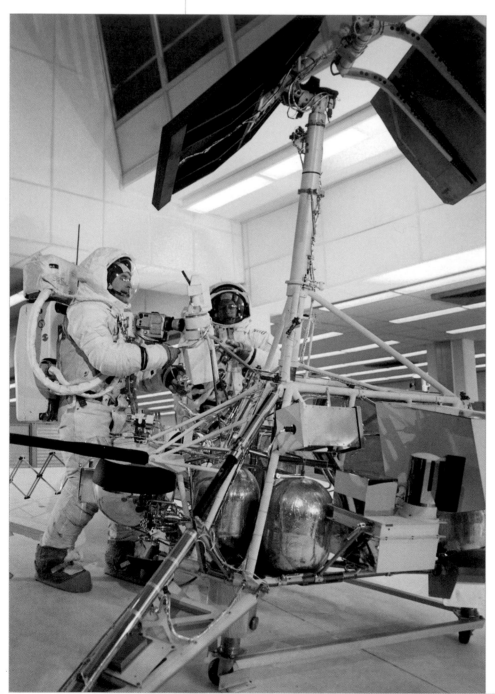

Five weeks before launch, Bean (*left*) and Conrad rehearse their second moonwalk with a mockup of the Surveyor 3 probe they will visit on the moon's Ocean of Storms.

Package, or ALSEP. The second walk would be an extended geologic traverse, culminating with a visit to Surveyor 3. Conrad and Bean would cut off pieces of the probe so the engineers could see what had happened to it during its thirty-one months on the lunar surface.

Conrad was thrilled with his mission. If he couldn't make the first landing, this was the next best thing. He and Bean weren't simply going to get down in one piece, grab a few rocks, and take off. They had the first lunar surface operations plan, a timeline packed with good, useful work for the scientists. And if they really did manage to land next to the Surveyor, they would open the way for the pinpoint landings the geologists wanted.

As the months of training wore on, the simulator instructors at the Cape considered Conrad, Gordon, and Bean one of the sharpest, most competent teams that had ever trained for space. Conrad joked that they should solve NASA's crew-selection worries by volunteering to fly every mission from then on, rotating seats on each flight: next time it would be Gordon-Bean-Conrad, then Bean-Conrad-Gordon. . . .

By the day before launch, Conrad felt ready. The mission he had worked toward for seven long years was finally upon him. That night, he and Gordon and Bean were joined for dinner in the crew quarters by Tom Paine. Later, in the parking lot, Paine made a promise to Conrad: If some problem came up and he didn't get to land, Paine said, he would put Conrad and his crew on the very next mission—so they shouldn't do anything rash. Conrad thanked Paine for his kind offer and said good-bye. He was halfway up the stairs when he realized Paine had made the same promise to Neil Armstrong.

II: SHORE LEAVE

MONDAY, NOVEMBER 17
11:45 P.M., HOUSTON TIME
3 DAYS, 13 HOURS,
23 MINUTES MISSION
ELAPSED TIME

"Twenty-four hours from now, Beano, we're on our way down, pal. That's when I get nervous! Find that little muthuh! And I'll land it right side up!" Pete Conrad was anxious, and he wasn't trying to hide it. But right now, as he and his crew settled in for their first day in lunar orbit, the atmosphere inside *Yankee Clipper* belied his tension. In the background, Frank Sinatra crooned "The Girl from Ipanema" on the tape recorder. The conversation was decidedly relaxed, unlike the sparse exchanges of Armstrong and his crew. It was the banter of three friends on vacation together in the wilderness.

"If they made up a Hollywood movie, just like this, you wouldn't believe it," said Bean.

"What do you mean?" asked Conrad.

"Listening to this music on the back side of the moon."

Gordon countered, "You got something against music?"

"Nobody would buy it," Bean said. "This is cornball—you gotta be *hard* out there."

"Say, the biggest thing I missed on Gemini 5 was not having any music . . ." said Conrad. What he would have given for that tape player then, cooped up in that tiny cabin for 8 days! For Apollo 12 he'd brought along some of his favorite tunes, mostly country-and-western stuff (like Bob Wills and the Texas Playboys' version of "San Antonio Rose") that sent Gordon and Bean into hiding. Bean's tape was Top 40, which they all liked well enough—especially the bubble-gum hit called "Sugar Sugar." When it came on during the trip out from earth, the three of them would hold onto the struts in the command module and bounce weightlessly to the beat, dancing their way to the moon. Conrad would say it emphatically in the debriefing: people have got to have some entertainment on these flights; you can't just look out the window. ☾

"Yeah," Conrad laughed. "Let's take the LM down and land on the back side. Wouldn't that shake 'em up?"

Conrad was happy with the way things were going. Since the lightning strike, they hadn't had a single problem worth mentioning. That was the best way to start a flight—get all the trouble out of the way early. And next to Gemini, the command module was like a hotel. There was warm water for rehydrating the food and the coffee; there was toothpaste, and shaving. By now, Conrad was looking forward to getting cleaned up. "Hey, I'm gonna take a bath tonight," he announced. "Take a bath, shave, get all cleaned up, good night's sleep—" He turned to Bean. "You got anything else to do tomorrow? All right, that's what we'll do then. We'll go for a little lunar landing, how's that? Unless you got something better in mind—a little surfing at the beach, or something."

"Hell, yes," Bean said.

Gordon joined in. "How about the back-side sand? Go play in the sand on the back side."

"*Yeah,*" Conrad laughed. "Let's take the LM down and land on the back side. Wouldn't that shake 'em up?"

But behind the banter, Conrad was ever mindful of what was coming up. No doubt about it, the stakes had changed with this mission. Just getting the LM down in one piece wasn't good enough. The real test was finding Surveyor 3.

According to the scientists, the Surveyor was sitting on the slopes of a worn, old crater 656 feet across. Immediately around it were several other craters, a bit smaller and sharper. It was this clump of craters that Pete Conrad would have to locate among thousands of others when his lunar module pitched over for its descent to the moon. Seen from the east, the view Conrad would have during his approach, the craters looked like the outline of a snowman, with the Surveyor sitting in the pit of its fat belly; he simply named it the Surveyor crater. He named the other craters appropriately: Head, Left Foot, Right Foot. On the photographs, there appeared to be a fairly smooth area on the near-right side of the Surveyor crater, and that was where Conrad would try to land. If that place—which became known as Pete's Parking Lot—turned out to be unsuitable, or if Conrad couldn't get to it, he must try to land close enough to the probe for him and Bean to reach it on foot without difficulty.

Before the flight, Chris Kraft had told Conrad not to stress the Surveyor when he talked to the press; otherwise if he landed off target the press would say the mission had failed, even if he and Bean accomplished all the other objectives.

Using shaving cream squeezed from a tube, Gordon shaves while Conrad looks on. This luxury had debuted on Apollo 10 after it became clear that there would be no stray whiskers to foul command module electronics.

But to Conrad, the bottom line was that if he and Bean didn't find Surveyor 3, all their planning would be for naught. The geologic traverse was designed around the various craters of the Snowman. And the future exploration of the moon depended on the pinpoint landing capability. It was up to mission control to put his lunar module on target, and it was up to Conrad to land.

TUESDAY, NOVEMBER 18
4:00 P.M., HOUSTON TIME
4 DAYS, 5 HOURS,
38 MINUTES MISSION
ELAPSED TIME

"I'm about as jumpy as I can be this morning." For Alan Bean, the anticipation was over; now it was time for him to do the thing he had been training to do all these months. Stick to the checklist. Be steady. Don't throw the wrong switch. Bean couldn't deny it; this was the big day.

"Oh, you noticed?" Conrad said wryly, as the two men ate breakfast over the lunar far side. Neil Armstrong had never said a word about anxiety, but Conrad wasn't Armstrong. "I just hope we find the old Snowman! Then I hope we find a place to land! Then I hope I can set it down all right!" Between bites of Canadian bacon, Conrad confessed, "It's driving me buggy. I just don't know what I'm gonna see when I pitch over. You know"—Conrad laughed at the thought of his own helplessness—"I'm either gonna say,

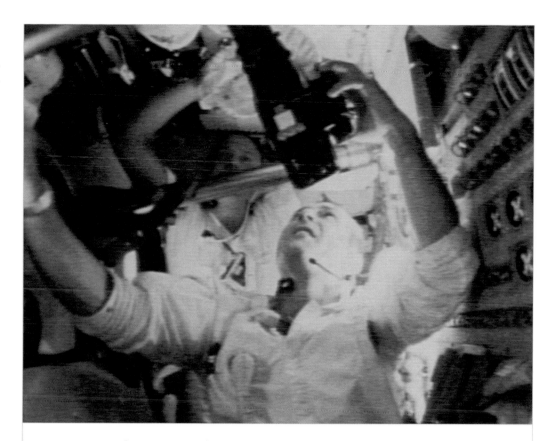

Wielding a Hasselblad camera, Gordon photographs the moon through *Yankee Clipper*'s hatch window while in orbit 69 miles above the moon. In a matter of hours, he will say farewell to Conrad, visible in the background, and Bean as they depart for the moon's surface.

Aaaaaaa! There it is! or I'm gonna say, *Freeze it, I don't recognize nothin'!*"

"If you don't recognize a thing," Bean offered, "just tell me. I'll look out my side, and you look at the computer for a few seconds, and let me see if I see anything out there."

Below them, the sun cast long shadows over a battlefield of craters. The near side of the moon didn't impress Pete Conrad, but the far side, with its enormous bumps and hollows—that did amaze him. He looked down at what seemed to be a string of small volcanoes. "That's fantastic," he exclaimed to Bean. Then he thought for a moment about where they were, and he said quietly, "How'd we ever get here anyway?" And he and Bean laughed. ☾

●◗◖◯◯◯◗◖

It had always amazed Alan Bean that in all the time they had spent training together, Dick Gordon had never once showed the disappointment he had to feel every time he heard Bean introduced as the guy who was going to the moon with Pete Conrad. There was never a trace of sarcasm from Gordon, never an ironic remark. He had done everything to make Bean feel welcome on the crew. Bean knew that if the situation had been reversed he could not have handled it as well. Now, Gordon was suspended at the end of the docking tunnel between *Yankee Clipper* and *Intrepid,* looking down at Conrad and Bean. It was a moment for good-byes, and yet none were said.

"I guess I've gotta close the hatch now," Gordon said. And he looked at his two friends for a moment, and then sealed the tunnel. Bean wondered whether he would ever see Gordon again.

Minutes later Gordon flipped the switch to release them, and for a few minutes *Yankee Clipper* and *Intrepid* flew in formation while Gordon took pictures. Then, as Conrad and Bean watched, Gordon pulled away.

Gordon kept the spindly, four-legged craft in sight through the 28-power sextant. When Conrad and Bean fired their descent engine to drop out of lunar orbit Gordon was looking straight up *Intrepid*'s engine bell. The spacecraft carrying his friends seemed to become a glowing ball, like the view into a jet engine climbing on afterburner. The glow lasted just under half a minute, then died.

WEDNESDAY, NOVEMBER 19

12:42 A.M., HOUSTON TIME

4 DAYS, 14 HOURS,

20 MINUTES MISSION

ELAPSED TIME

If the lunar module belonged to any astronaut, it was Pete Conrad's. He knew everything about the LM cabin because he'd helped to design it. Back in 1963 he'd been in on some of the major changes—taking out the seats and letting the pilots fly standing up; replacing the round, forward hatch with a bigger, square one that could accommodate an astronaut wearing a bulky backpack; adding a small rendezvous window above the commander's head (*that* had been a battle; they'd tried to tell him, the pilot, that he didn't need one), and the list went on. As *Intrepid*'s descent engine ignited 50,000 feet above the moon, Conrad suddenly flashed back to five years earlier, when he was visiting the Grumman plant on Long Island, and standing in a mockup of the LM, then known as "the bug." The instrument panels were nothing but drawings pasted onto plywood, but Conrad had stood there and imagined himself descending to the surface of the moon. Now it was happening. *Intrepid*'s engine came silently to life, right on schedule. Bean welcomed the acceleration; after five days of weightlessness it felt good to be *standing* again. Conrad scanned the gauges and *Intrepid* continued its long ride down.

Ever since the pinpoint landing became his mission last summer, Conrad had wondered whether Kraft's people really could pull it off. But they were confident—so confident, in fact, that Conrad sometimes had trouble taking them seriously. One day Conrad was talking to trajectory specialist Dave Reed about his landing point. Reed asked, "Where do you want me to put you?" Conrad doubted it would make much difference what he told Reed—after all, Armstrong and Aldrin had landed four *miles* off target—but Reed was talking like a travel agent making an airline reservation. Conrad went along and picked a spot, with due consideration to sun angle, traverse distance, and the like. Then after a few simulated landings, he changed his mind and went back to Reed for a new spot a few hundred feet to one side. Without batting an eye, Reed set about figuring out the necessary changes to the software, and Conrad couldn't believe the man was serious. He blurted, "You can't hit it anyhow! Target me for the center of the Surveyor crater." Reed answered, "You got it, babe." But Conrad would believe it when he saw it. ❦

12:51 A.M.

At 8,000 feet, with pitchover almost upon them, Conrad leaned forward in his space suit, straining to glimpse the Snowman. He told Bean, "I think I see my crater. . . . I'm not sure. . . ." Then, right on schedule at 7,000 feet, *Intrepid* pitched forward for the final descent. Suddenly Conrad beheld a ghostly, black-and-white panorama, seemingly too stark to be real—ten thousand shadows inside ten thousand craters. Conrad felt a stab of unease: he couldn't find a single feature he recognized. But when he sighted along the 40 degree mark on his window, suddenly—"Hey, *there it is.* There it is! Son of a gun, *right down the middle of the road!*" Far in the distance, Conrad could see the Snowman, tiny, almost lost in the sea of craters; *Intrepid* was headed for the very center of the Surveyor crater. Even as Bean tried to read him LPD angles, Conrad was too excited to listen—"Look out there! I can't believe it! Fantastic!" But the targeting was almost too good. Conrad wasn't about to land in the middle of the Surveyor crater. As the computer flew *Intrepid* down, Conrad told it to shift the aim point short and to the right of the crater. He would give Pete's Parking Lot a try after all. He heard Jerry Carr's voice: "*Intrepid,* Houston. Go for landing."

A thousand feet above the moon, Bean stole a glance out his window and saw a bright field of craters, with the Surveyor crater as big as life, directly ahead. They were coming in at great speed. Up to now, the landing had seemed like a simulation, but this was almost more than he could stand to look at. It was amazing, even frightening. Quickly, he turned his gaze back to the instruments. His voice was calm as he told Conrad, "Looks good out there, babe; looks good."

But Pete's Parking Lot didn't look very good to Conrad; it looked more like Pete's battlefield. Again he shifted the landing point, this time further downrange.

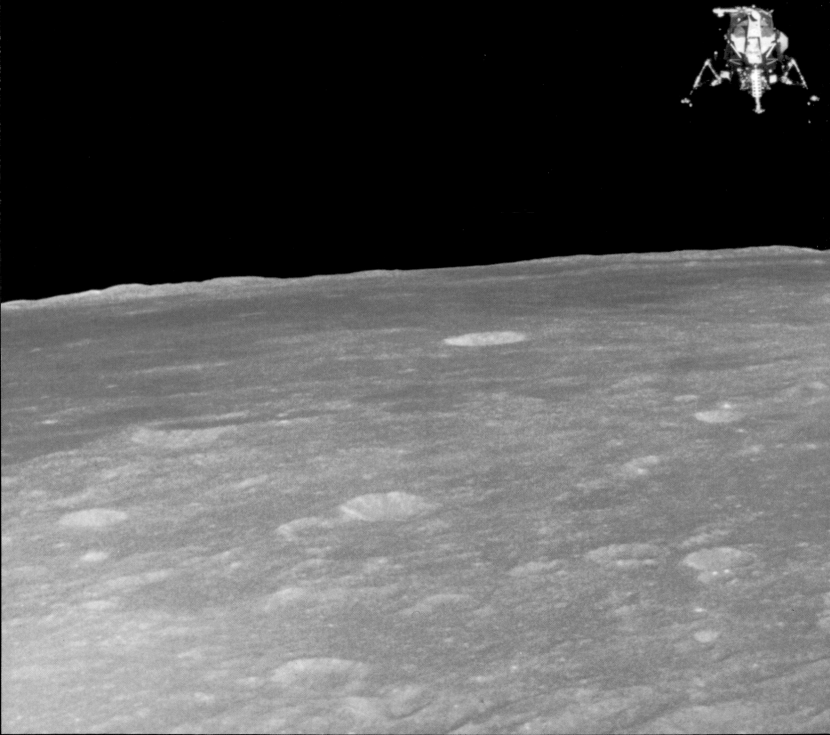

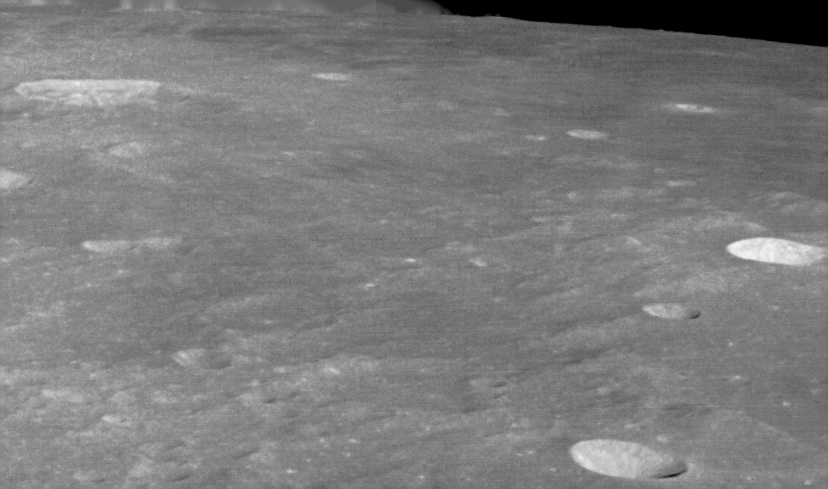

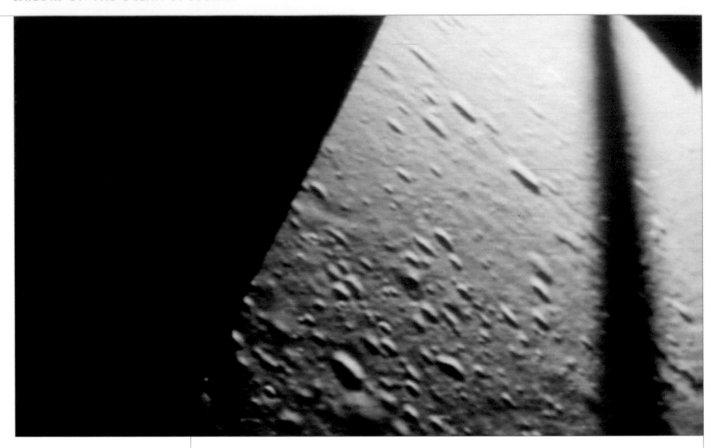

Pete Conrad's first view of his landing site, recorded by *Intrepid*'s movie camera, was a confusing jumble of craters. The dark stripe at right is the silhouette of a safety bar attached to the LM window.

"You're at five hundred and thirty feet, Pete," called Bean. "You're all right!" But they were still coming in like a bullet. Conrad wanted time to slow down and look for a landing spot, and he could afford to; they were fat on fuel. As *Intrepid* descended through 400 feet, Conrad took over. He pitched the craft back to kill their forward speed. *Intrepid* slowed, but not as soon as Conrad wanted. He saw the Surveyor crater drift past and said, "Gosh, I flew by it." Conrad looked out to his left, just beyond the crater's northwest rim, and saw where he wanted to land: a smooth area between Surveyor crater and Head crater. He banked *Intrepid* hard to the left. The craft responded with the same sluggishness as the LLTV at Ellington. Bean, who had never flown the landing trainer, was surprised to see the 8-ball in front of him tilt so sharply. He wondered what Conrad was doing. He said in a calm Texas drawl, "Hey, you're really maneuverin' around."

"Yep," Conrad answered, too busy to say anything more.

"Come on down, Pete," Bean coached, his eye on the rate-of-descent gauge. Bean knew that Conrad must not descend too slowly while they were this high up; it would cost too much fuel. "Two hundred feet, coming down at three," Bean said. "Need to come on down."

"Okay," Conrad answered. In the simulator, he had always waited until they were down to 100 feet or so before arresting the last bit of forward speed and starting the final, vertical descent. But some instinct now prompted him to level off early. It was the best decision he made, because within moments, his view of the surface began to blur. *Intrepid* was kicking up a tremendous amount of dust, far more than Armstrong and Aldrin's movies had suggested. The dust shot away from him in bright streaks, rushing to the horizon.

"Ninety-six feet, coming down at six. Slow down the descent rate," Bean called quickly, urgently.

The dust blanket thickened. Conrad could see nothing but streaks, with a few rocks sticking up here and there. He had no idea whether there were any craters directly underneath him, but he would have to take his chances.

"Lookin' real good," said Bean, his eyes glued to the gauges. "Fifty feet, comin' down. Watch for the dust." Bean didn't see the dust storm already raging outside his window.

Still *Intrepid* crept downward. Conrad's eyes flicked back and forth between the window and the instruments. It was absolutely the worst way to have to fly, with his attention split, but he had no choice. The gauge that was supposed to display lateral motion seemed to be broken, forcing him to look outside to make sure he wasn't drifting sideways or, even worse, backwards. ☾

"Thirty feet, coming down at two," Bean said. "Plenty of gas, babe, plenty of gas. . . ."

"Thirty seconds," warned Jerry Carr in mission control. Bean answered, "He's got it made!"

Conrad was now flying entirely by instruments. He had planned to wait until touchdown to shut off the engine, but suddenly, he saw the blue glow of the Contact Light and instinctively his hand went to the ENGINE STOP button, and *Intrepid* fell the last few feet to the dust with a firm thump. Conrad had logged a hundred seconds of stick time—the only real flying he would get on Apollo 12—but they had required every ounce of experience from twenty years of piloting. He had landed on target; he was sure of that. And just now his friend and former student was slapping him on the back, saying, "Good landing, Pete! Out-*stand*-ing, man!"

"You're headed right square out the hatch. Wait, wait, wait—Come forward a little. There you are." With Bean giving him guidance, a fully suited Pete Conrad crawled out of *Intrepid*'s front hatch and onto the porch. Caution dictated that he move slowly in the lunar vacuum, but Conrad was impatient. From

Caught by Al Bean's camera, Conrad pauses at the top of *Intrepid*'s ladder before he descends to the bright, dusty Ocean of Storms. At right is part of the lunar module's open hatch.

the moment of touchdown he couldn't wait to get outside; he had to get down to the surface and find out once and for all whether the Surveyor was really there. Looking out the window it was impossible to tell where they were. Like Armstrong and Aldrin, he and Bean couldn't gauge distances or sizes, and that made it all but impossible to figure out what craters were in front of them. But when Conrad finally stood on the top rung of the ladder, he could actually see part of the Surveyor crater—in fact, he realized with a mixture of relief and alarm, he had landed *Intrepid* not ten yards from its rim. He headed down the ladder while a color TV camera broadcast the scene to earth. Conrad knew people wouldn't remember the third man to walk on the moon; there was no need to make up something momentous to say. But he did have a quote; in fact, he had a bet to win.

The bet had its origins in the heat of a Houston summer afternoon when Conrad and his wife, Jane, were by the pool, entertaining Italian journalist Oriana Fallaci. Conrad had known Fallaci since 1964, when she came to Houston to write a book on the astronauts. Fallaci never had any trouble speaking her mind, especially when it came to bureaucracy. And that afternoon she was convinced that NASA's bureaucrats had told Neil Armstrong what to say when he stepped on the moon. Conrad tried to convince her otherwise, but she was certain of it. Conrad persisted; he couldn't swear that Armstrong had written "One small step for a man, one giant leap for mankind," but he was sure that whoever had penned it, Armstrong had chosen what he would say. "Look," he told his guest, "I'll prove it to you. I'll make up my first words on the moon right here and now."

"Impossible," Fallaci pronounced in her thick accent. "They'll never let you get away with it."

"They won't have anything to say about it, Oriana. They won't know about it until I'm on the moon." Conrad had a good idea: Since he was nearly the shortest guy in the Astronaut Office, why not say . . .

When Fallaci responded, "You'll never do it," Conrad answered, "How about five hundred bucks?" They shook on it.

Conrad reached the last rung of the ladder, held on with both hands, and jumped. He fell to the footpad with a gentle bump.

"Whoopie! Man, that may have been a small one for Neil, but it's a long one for me." ☾

As Neil Armstrong had done, Conrad held on with his right hand and placed his left boot on the moon. He swirled his foot in the dust. "Ooh, is that soft and queasy. . . . I don't sink in too far. . . ." Still holding on, he planted both feet on the surface, then let go. His first steps brought him into the blinding glare of the sun. He turned away from it, moving in small, floating steps across the shadowed ground. Leaning forward at an impossible angle to compensate for his backpack, he felt as if he might fall over at any moment.

"Well, I can walk—pretty well—ah—Al, but I've got to watch where I'm going." Conrad kept walking until he could look past *Intrepid*'s foil-covered bulk to the east, across a vast bowl twice the size of a football stadium, and he could hardly believe what he saw. On the crater's shadowed far wall sat a tiny, spindly white shape: Surveyor 3. He let out a high-pitched cackle. "Does that look neat! It can't be any further than six hundred feet from here. How about that?"

In Houston, Capcom Ed Gibson radioed congratulations, but his words were drowned out by the sound of applause. The trajectory people had done it. Conrad was elated.

But the trip to the Surveyor would wait until tomorrow. Right now, Conrad had work to do. And after a few minutes, everyone listening in Houston

Some 600 feet behind *Intrepid* in this picture, taken during the first Apollo 12 moonwalk, stands the dim shape of Surveyor 3. This unmanned probe, which reached the moon in 1967, was the target for the first pinpoint lunar landing.

SURVEYOR 3

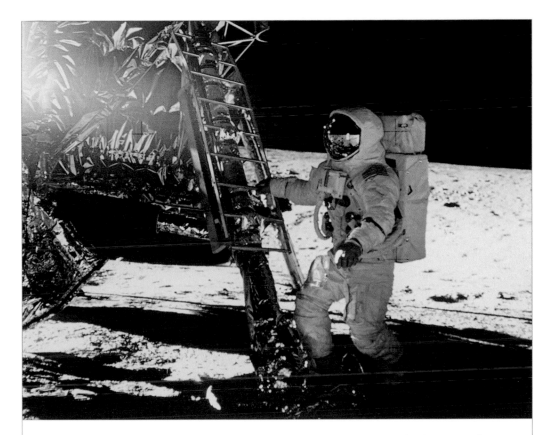

The fourth man to walk on the moon, Al Bean steps into the lunar dust. At this unique instant in his life, Bean later recalled, he focused his attention on the many tasks scheduled for the four-hour moonwalk.

and around the world heard an unaccustomed sound from the lunar surface: the sound of Pete Conrad humming. *"Dum de dum-dum-dum . . ."* Nothing in particular, just tuneless, mindless humming. *"Dum diddee dum-dum-dum . . ."*

Inside *Intrepid,* Bean didn't notice the humming at first. He'd heard his commander do the same thing in the simulator, and flying a T-38. Even for Pete, though, this was more than usual. Every few minutes the tune that had no tune crept onto the earth-moon airwaves. In the press room at the Manned Spacecraft Center, the reporters covering the flight half-jokingly wondered if Conrad was on an oxygen high. But to more accustomed ears, this was simply the sound of Pete Conrad in his no-sweat mode. None of his worries had been realized. And now Conrad saw with delight that it was even easier to do his work in one-sixth g than he'd expected. If everything kept going this well, the mission—not to mention some lunar rocks and a few pieces of Surveyor 3—was practically in the bag.

6:13 A.M.

For half an hour, Alan Bean had watched as his commander danced over the craters, 20 feet below. Now he would follow. As Conrad gave directions, Bean inched backward through the hatch and descended the ladder. When he

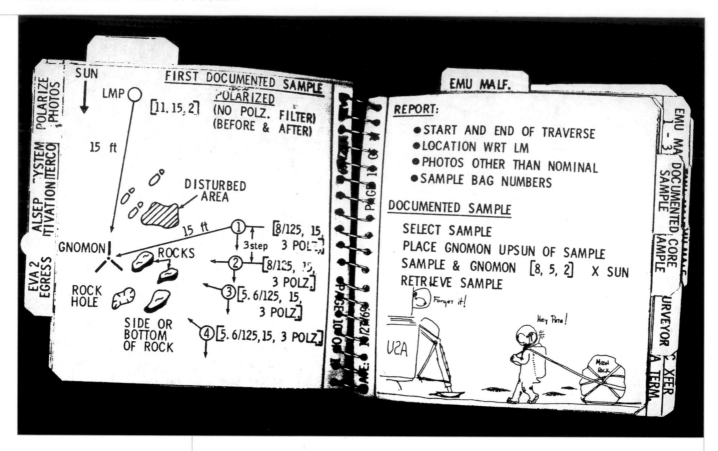

Each astronaut had a checklist, like this one of Conrad's, attached at his wrist to remind him of tasks and procedures for each moonwalk—and sometimes offer a laugh or two. The cartoon of Bean, as a Snoopy-like character, retrieving an impossibly large moon rock was drawn by one of the Apollo 12 backup crew.

reached the footpad Conrad was waiting with his camera: "Okay, turn around and give me a big smile. Atta boy. You look great. Welcome aboard." For the rest of his life, people would ask Bean what he was thinking at this moment, and he would tell them that his thoughts were on one thing: his checklist. On his left cuff, he wore a checklist that looked like a small spiral notebook; Conrad had one too. On those pages, almost every minute of their time on the moon was accounted for. Before the flight it had been Bean, with his love for detail, who had delved into the task of helping to write that checklist. But before he set about his assigned tasks, there was something he wanted to do. Reaching into a storage pocket on his thigh, he pulled out his silver astronaut pin, the one he had worn on his lapel during his six years as a rookie. When he returned to earth, he would receive the gold pin of a space veteran. Now, with a few halting steps, Bean stood at the edge of the Surveyor crater; he flung the pin into it.

According to the checklist, Bean had 5 minutes to gain his balance and learn to walk. He was amazed at his new buoyancy: "You can jump up in the air—"

"Hustle, boy, hustle! We've got a lot of work to do." Conrad's voice. Conrad had one rule: Stay on the timeline. It was true that in their two moonwalks the two men would spend more than 7 hours outside, about three times the amount Armstrong and Aldrin had, but all of that time was packed with objectives, and Conrad did not want to fall behind. Right now, he was setting up an umbrella-shaped S-band antenna to improve communications with earth. He called to Bean, "How about doing some useful work, like getting that TV camera going."

"Okay," Bean answered brightly, "good idea. Let's get that TV out and show everybody." That didn't go as planned; while Bean was carrying the camera on its tripod away from the LM, he accidentally pointed it too close to the sun for a few long seconds. By the time Bean realized the mistake it was too late. The moonscape on televisions in Houston and across the world had changed into a jumble of black-and-white shapes. Nothing Bean tried, even rapping on the camera with his geology hammer, seemed to work. History's second moonwalk had become a radio show—but with Pete Conrad on the moon, who needed television? When he wasn't talking to Bean, he was talking to himself, or he was humming, or laughing. In the midst of some task he would suddenly cackle, and no one knew why—except the backup crew, Dave Scott and Jim Irwin, who had arranged for a few . . . *additions* to the cuff checklist. They had adorned the pages with cartoons of Conrad and Bean as Snoopy astronauts, just like the comic-strip beagle's space-faring persona. But what really made Conrad laugh were the *Playboy* pinups, reduced to about three inches square, annotated, of course, with proper geologic terminology: "Don't forget: Describe the protuberances. . . ."

The funny thing was, there *were* a couple of mounds poking up from the undulating plain. The geologists had asked him to keep an eye out for unusu-

During their first moonwalk, Conrad and Bean encountered this geologically puzzling mound of lunar soil not far from *Intrepid*. Scientists later theorized that it and another mound nearby were pushed up by rocks thrown out of a crater.

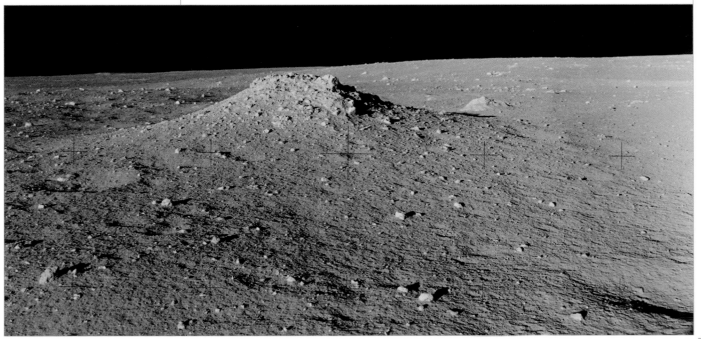

al features, and here they were. They were a few hundred feet from the LM; Conrad made a mental note to get over and take a look at them, perhaps after the ALSEP work.

"We're making our move, Houston." As Conrad announced their departure, Bean lifted what looked like a strange, square set of barbells and began walking west. The barbells were actually two pallets loaded with the ALSEP's scientific instruments. There was a seismometer to measure moonquakes, a magnetometer to look for a lunar magnetic field, another sensor to sniff out the moon's incredibly tenuous atmosphere, and others to search for ions in the moon's vicinity and analyze high-energy subatomic particles emanating from the sun. There was also a central transmitting station to relay the ALSEP's data to earth. Together, these experiments comprised the first full-fledged scientific station to be set up on another world. ☾

Bean struggles to pull a plutonium fuel element from a graphite-lined case on *Intrepid*'s descent stage. The fuel element, necessary to power the mission's package of scientific experiments, refused to budge —until Conrad whacked the case with his geology hammer.

Bean planned to carry the ALSEP several hundred feet from the LM, far enough so that the instruments would not be affected in any way by the dust kicked up when he and Conrad blasted off tomorrow. As he walked, Bean could feel his heart pounding with the effort to grip the bar with his stiff, pressurized gloves and hold the packages out in front of him. "I'm going to set it down and rest," he said. Meanwhile, Conrad ran out ahead of him, scouting a good area to lay out the experiments.

In their 2½ hours on the Sea of Tranquillity, Neil Armstrong and Buzz Aldrin had never ventured more than a couple of hundred feet from their lander. Now, 1 hour and 48 minutes into this first moonwalk, Conrad and Bean were extending the reach of lunar exploration more than twice over. Bean walked until he came to a level place about 500 feet from the LM.

Bean couldn't count the number of times he and Conrad had practiced setting up the ALSEP, on a simulated lunar surface behind

the Manned Spacecraft Operations Building. To an unaccustomed eye the experiments looked like strange, delicate creations. In truth each was an engineering marvel, built to withstand lunar heat and cold and to perform on a minimum of electrical power—and be small and lightweight to boot. Bean knew each of them well. There was the seismometer, which looked like a silver paint can atop a round, silver drop cloth. And the magnetometer, with its three, gold-foil-tipped arms reaching into the vacuum. The small ion-detection experiment, with its ridiculously short legs joined by a spider web of wires. And the squat little solar wind spectrometer with its odd facets and bubbles. Each of them would make incredibly sensitive measurements to probe the moon and the void around it.

Bean could feel his heart pounding with the effort to grip the bar with his stiff, pressurized gloves and hold the packages out in front of him.

Laying out the ALSEP wasn't the kind of work people expected an astronaut to be doing on the moon; it was more like arranging garden furniture than like exploration: Undo bolts. Set each experiment on the ground. Tamp the dirt and make the instrument level. Make sure each one is pointed in the proper direction with respect to the sun, using a shadow indicator, and that each is the proper distance from the others. And—keep them from getting dirty. This was the job, and as he worked, Bean felt pride of accomplishment.

There was Conrad, a few yards away, working on the central station. His white suit glowed in the sunlight, except from the knees down, where it looked as if it had been dragged through a coal bin. Bean knew that Conrad was not immune to awe; you had only to be in the same spacecraft with him orbiting the moon. He'd talked a blue streak about the craters and mountains and lava formations and how amazing it all was. The important thing was, it never slowed him down. He was always spring-loaded, ready for the next event. Around him, the Ocean of Storms stretched in all directions, an undulating plain that rose and fell with craters of every size. Surprisingly, if Bean didn't think about it, the moon didn't really seem like an otherworldly place. But if he looked high into the black sky, he saw the earth, and the sight of it was enough to bring home the electrifying realization of where he was. And

Like a weightlifter walking with barbells, Bean deploys the science experiment packages. As he walked he could feel heat from the plutonium fuel element, located near his right hand, penetrate his insulated space suit.

he told himself, *You don't have time now; you can think about this after the flight.* But just now, things were going well, and when he began to deploy the magnetometer Bean allowed himself a moment of fun. Removing a set of Styrofoam packing blocks, he took one of them and gave it a sidearm fling, and it sailed into the distance for an impossibly long time, tumbling in slow motion against the black sky, before landing. He took a second block and called out to Conrad, who was busy working.

"Pete? Pete!"

"What?"

"Watch this." Bean made an underhand toss. The white shape ascended on a long, lazy arc and hit the dust. "Boing!"

"Stop playing," Conrad laughed, "and get to work. Come on, maybe they'll extend us until four and a half hours. I feel like I could stay out here all day." It was true; they could have handled a moonwalk twice as long as the 3½ hours allotted, and that's what Conrad would say in the debriefing when he was back on earth.

Conrad prepares to set up the Lunar Surface Magnetometer, the object near the center of the picture with tubes wrapped in gold foil. The purpose of the experiment was to gauge the strength and orientation of the moon's magnetic field.

It took Conrad and Bean over an hour to set up the ALSEP, but when they were done it resembled an odd, five-pointed star with the central station in the middle and the experiments radiating from it on bright orange ribbons of cable. They looked just like the ones in training—except for one problem: It was almost impossible to walk past an experiment without spraying dirt on it. In no time the clean white surfaces were sprinkled with black powder. There wasn't any point in trying to wipe it off; that would only have ground it in.

"I remember how they took care of this white paint," Bean said. "You had to have gloves to touch it. Remember?"

Conrad laughed. "Yeah, they got kind of a problem here."

Well, Bean decided, there wasn't any point in worrying. Dirty or not, the experiments would just have to work. Finally the ALSEP was laid out and ready, and Conrad activated the central station. At that moment, in a back room down the hall from mission control, scientists gathered excitedly around a set of chart recorders. On the tracing for the seismometer, they saw little squiggles and bumps—not from moonquakes but from Conrad's and Bean's footfalls as they headed away from the ALSEP to gather samples.

IN LUNAR ORBIT, ABOARD *YANKEE CLIPPER*

Every two hours a tiny, unblinking star appeared in the eastern sky above *Intrepid*, ascended to the zenith, then vanished in the west. It was bright enough for Pete Conrad to see with the naked eye from the LM cabin. The star was *Yankee Clipper*, carrying Dick Gordon on a 38-hour solo voyage around the moon. Before the mission, Gordon had wondered whether he would think of the earth and feel his awesome separation from humanity. Instead, he savored his aloneness—and mostly he was too busy to think about anything besides the flight plan. But when there was a lull, his thoughts turned to his friends on the Ocean of Storms. Gordon envied Conrad and Bean more than he could say. Mike Collins had said he was perfectly happy going 99.9 percent of the way, but that wasn't how Gordon felt. Walking on the moon, not orbiting it, was the name of the game; it was as simple as that, and it had been his goal from the day he joined the astronaut corps. Gordon rationalized his position, telling himself that down the line, on one of the later missions, he would get his chance. But when he watched Conrad and Bean climb into that lunar module and pull away, he wished he were going with them. His crewmates knew it; Conrad had said he wished the LM held three people so they could all land together. When Conrad steered the lander to a touchdown Gordon was listening, and as the dust settled on the Ocean of Storms he radioed congratulations and told his friends, "Have a ball."

On the next orbit, when he flew over the landing site, he was ready to try to spot *Intrepid* on the surface. Unlike Mike Collins, Gordon knew exactly where to look, and he knew the navigation was so good that all he had to do was tell the computer to point the sextant right at the Snowman. At the proper moment the optics whirred into position and there, just off the rim of the Surveyor crater, he saw a point of light with a needlelike shadow, and he knew he had found them. "I have *Intrepid*," he announced to Ed Gibson, and as he got closer, he could almost convince himself he could see details, four landing legs sticking out from the descent stage, though he knew the sextant wasn't powerful enough for that. He was in the middle of reporting *Intrepid*'s

Dick Gordon trains in the command module simulator for his 33 hours alone in lunar orbit. During the real thing, he decided to remove his cumbersome space suit, trusting in *Yankee Clipper*'s hatch seal to preserve the five pounds of atmospheric pressure he needed to survive.

position when he spotted another point of light nearby: "I see Surveyor! I see Surveyor!" Gordon was jumping up and down, he would say later—if it's possible to do that in zero g. If he could not make the first pinpoint lunar landing, then he had been the one to discover that it had been a success. He told Gibson, "That's almost as good as being there." ☾

When Conrad and Bean took their moonwalk, Gordon was there vicariously. Thanks to mission control's relay he could listen to his friends as they bounced along. At one point, as the two moonwalkers wrestled with a balky piece of gear, Gordon couldn't help but laugh at what he was hearing. Later, he would tell his crewmates, "You were raising hell about some stupid device, and I was laughing my ass off."

In the moonwalk's last hour, Gordon prepared for his most important task of the day, a burn of *Yankee Clipper*'s SPS engine to adjust his orbit. If you could stand on a fixed platform in space, you would see that the moon slowly turns as it orbits, at exactly the rate needed to keep the near side pointed at earth. You would also notice that the orbit of *Yankee Clipper* stayed in a fixed orientation, with the result that each new circuit of the moon brought Dick Gordon slightly west of his previous position. If he did nothing, then over the 31½ hours that Conrad and Bean were on the moon *Yankee Clipper*'s orbit would shift far enough from the landing site to ruin the prospects for tomorrow's rendezvous. This 14-second firing would nudge *Yankee Clipper* into the proper orbit.

By now Gordon was very tired. Of course, test pilots are used to performing while fatigued, but he did *not* want to make a mistake with the whole world watching. But now, it was just him inside *Yankee Clipper*, surrounded by a forest of switches and dials, doing a task that had been performed by three men on every previous lunar mission. As he prepared for the burn he could hear Conrad and Bean going on excitedly about mounds and rocks, and the chatter was so incessant that it drowned out mission control, just when Ed Gibson was trying to read up crucial data on the burn. "If you're going to talk to me you're going to have to cut out the relay," Gordon told Houston. "It's impossible with those guys yacking." Gibson finally had to tell Conrad and Bean to be quiet for a few minutes. For safety's sake Gordon read his checklist over the air as he performed it, so that the flight controllers could check his work. Finally the engine lit and he was slammed into his couch for 14 long seconds. When it was over a relieved Gordon began to prepare for sleep. Without his crewmates to help out, everything took longer than usual. It was another two hours before he finished dinner and the lengthy cleanup for the night. He thought, Now's the time to think up all sorts of fancy prose to tell people what it's like to be the lonesome man up here. In truth, he was so tired he was glad just to go to sleep. ☾

III: IN THE BELLY OF THE SNOWMAN

**ON THE OCEAN OF STORMS
5 DAYS, 5 HOURS,
25 MINUTES MISSION
ELAPSED TIME**

Wide awake, Alan Bean lay in his space suit on a Beta-cloth hammock that was strung to the sides of *Intrepid*'s cabin. Immediately above him, lying fore-aft, Pete Conrad was asleep in his own hammock. Bean looked at his watch. It was 3:45 in the afternoon, Houston time; he and Conrad were halfway into a planned nine-hour sleep period. But sleep did not come easily to Bean. For one thing, his space suit was uncomfortable, even without helmet and gloves. He would have preferred to take it off, but that wasn't a good idea, not with all this dust; there was too much risk of clogging a zipper or a wrist ring. But the suit wasn't all that kept Bean awake.

Bean had always felt that he thought more about the risks of the job than other astronauts. Still, Bean knew, you can never be sure what goes on in

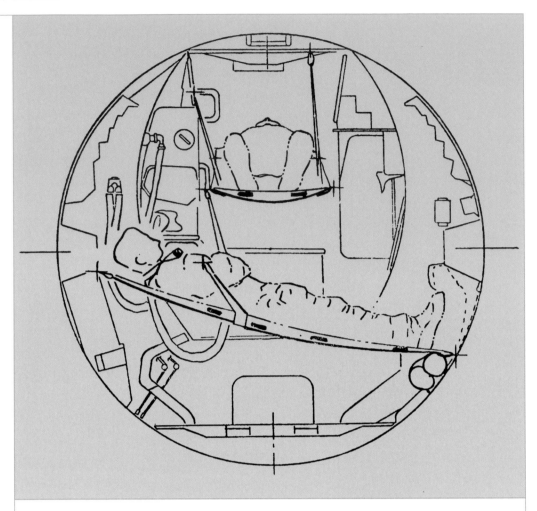

Intrepid was the first lunar lander equipped with hammocks for the moonwalkers to sleep in. Conrad occupied the upper hammock, strung fore-and-aft; Bean slept in the one hung side to side below Conrad's.

another man's mind, and it wasn't the kind of thing anybody ever talked about. It wasn't that he was anxious about anything in particular; it was more a heightened awareness of his surroundings. Bean heard the whine of *Intrepid*'s cooling pumps. A few hours ago, he and Conrad had just fallen asleep when one of the pumps changed pitch, and both of them awoke with a start, then went back to sleep. He looked up at Conrad's hammock. Even in a space suit Conrad's body barely weighed enough to sag the cloth; it lay almost flat in the lunar gravity. Before the flight Conrad had told him, "Don't worry, if anything goes wrong, it'll be something you've never seen before." The lightning strike proved him right. But that had not been a harrowing experience for Bean—although, he thought wryly, it would have been if he'd understood what was really happening. Pete had understood, and he'd been cool enough for all three of them. Of course, when it came to the landing on the moon, Conrad had done his share of worrying—that was only natural—and when he did, Bean was able to be reassuring, because he knew that if anybody could pull it off, Conrad could. Bean didn't worry much about the piloting end of Apollo 12; in his mind a mechanical problem was the thing to watch out for. All through the mission, even with the exhilaration of the ride, and the wonder of such incredible sights, Bean heard the background noise of his own awareness, reminding him that one of the rules of this game is that all machines eventually fail. For Bean, being on the moon was laced with that awareness.

Bean thought about the TV camera. He wondered why it didn't work. He had no way of knowing that the full-strength lunar sunlight had burned the light-sensitive coating right off the vidicon, that it was beyond hope. The camera should have been ready in time for them to train with it at the Cape, but all they'd had was a block of wood. All he knew was that somehow, he had managed to screw it up, but there was nothing he could do about it now.

Bean thought about the Surveyor. Over dinner, he and Conrad had talked about how the shadowed crater wall looked so steep they might not be able to walk along it safely. They would have to wait and see. For now, Bean knew, he had to get some rest, or his performance would suffer. If you're not doing something productive, Bean told himself, you should be sleeping. Finally, he drifted into a light slumber.

Meanwhile, in the top bunk, Pete Conrad soon awoke—not because his mind was too active, but because he was in pain. The right leg of his space suit had been misadjusted before the flight and was slightly too short; now, as he lay in his hammock, his suit bore down on his shoulder like a vise. He called down to Bean and told him they would have to do something about it before the moonwalk. They got up, took down the hammocks, and then, while Conrad sat on the ascent engine cover, Bean set to work. The leg of the suit was adjusted by a set of cords laced around the calf like sutures. Bean had to undo each cord, which was tightly knotted (because nobody wanted it to come undone), let it out a little, and retie it. The whole process took about an hour. Then Conrad called Houston, two hours ahead of schedule, and the two men began the new day. ☾

Morning lasts a week on the moon. For seven days, the sun slowly climbs in the black lunar sky, shrinking the shadows of rocks and craters. By lunar noon, temperatures climb to 225 degrees Fahrenheit. It takes another week for the sun to descend and then vanish below the western horizon. During the frigid two-week lunar night, temperatures plummet to 243 degrees below zero. Then the cycle begins anew.

A BRIEF HISTORY OF THE SPACE SUIT

The Apollo lunar space suit capped a long history of experiment and speculation. In 1934, aviation pioneer Wiley Post created the first true pressure suit for his flights into the stratosphere *(below)*. Pilots who flew the X-15 rocket plane to the edge of the atmosphere wore space suits as a precaution against loss of cabin pressure or an emergency ejection. An early moon-suit concept created by inventor Allyn Hazard featured a hard shell that was as cumbersome as it was protective. The suit that went to the moon *(right)* was essentially a wearable spacecraft. It is still praised by the Apollo astronauts who wore it.

MOON SUIT,
EARLY SIXTIES
CONCEPT

WILEY POST
PRESSURE SUIT

X-15
PRESSURE SUIT

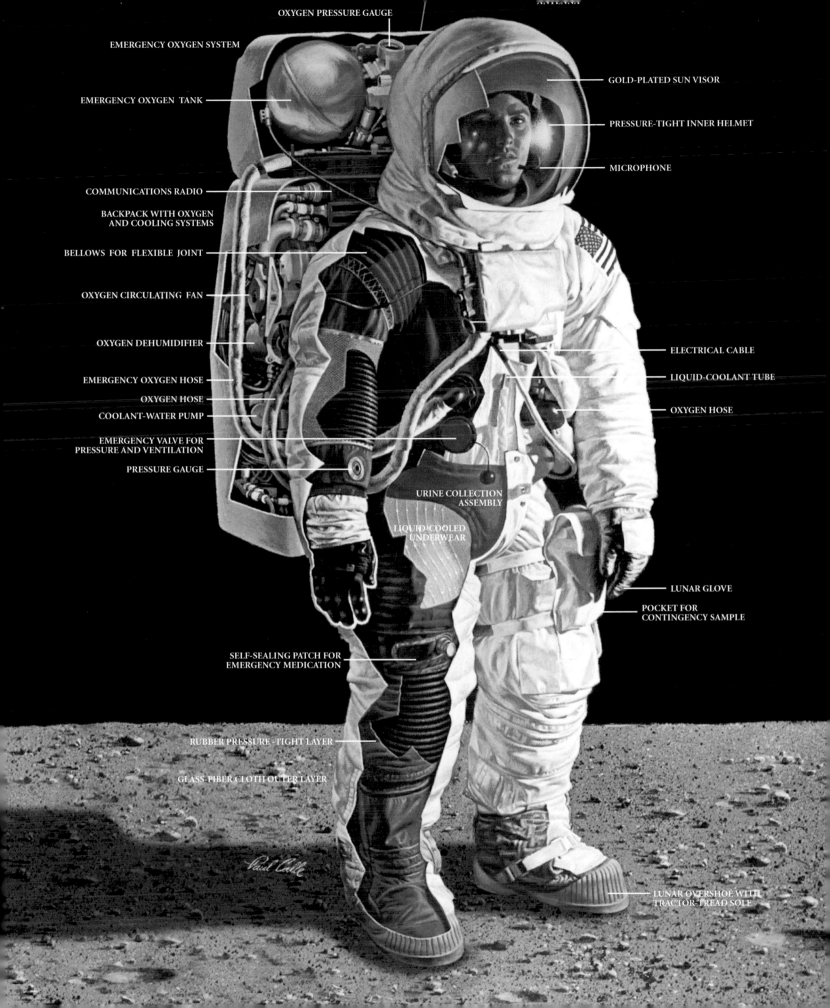

OXYGEN PRESSURE GAUGE

EMERGENCY OXYGEN SYSTEM

EMERGENCY OXYGEN TANK

COMMUNICATIONS RADIO

BACKPACK WITH OXYGEN
AND COOLING SYSTEMS

BELLOWS FOR FLEXIBLE JOINT

OXYGEN CIRCULATING FAN

OXYGEN DEHUMIDIFIER

EMERGENCY OXYGEN HOSE

OXYGEN HOSE

COOLANT-WATER PUMP

EMERGENCY VALVE FOR
PRESSURE AND VENTILATION

PRESSURE GAUGE

URINE COLLECTION
ASSEMBLY

LIQUID-COOLED
UNDERWEAR

SELF-SEALING PATCH FOR
EMERGENCY MEDICATION

RUBBER PRESSURE-TIGHT LAYER

GLASS-FIBER CLOTH OUTER LAYER

GOLD-PLATED SUN VISOR

PRESSURE-TIGHT INNER HELMET

MICROPHONE

ELECTRICAL CABLE

LIQUID-COOLANT TUBE

OXYGEN HOSE

LUNAR GLOVE

POCKET FOR
CONTINGENCY SAMPLE

LUNAR OVERSHOE WITH
TRACTOR-TREAD SOLE

When Conrad and Bean ended their first moonwalk it was about 6:30 A.M. local moon time. Nearly thirteen hours later, they stepped outside for the second moonwalk. But in lunar time only half an hour had elapsed. The sun had climbed only a few degrees, and yet everything looked slightly different. It almost seemed to Bean that a new landscape had taken shape while they slept. He kept noticing rocks that he thought he'd missed the day before; soon he realized they were the same rocks under different lighting. The colors of the surface—the spectrum of grays and tans and browns that changed as he looked in different directions—seemed a little more vivid. But the real change was in the Surveyor crater. No longer did its walls seem steep and forbidding; it had been transformed into a gentle bowl. Getting to the probe looked to be easy. Along the way, Conrad and Bean would test their skills as

Conrad had managed to smuggle something else in the pocket of his space suit, and when the time came, he and Bean would have their own little caper on the lunar surface.

lunar field geologists. For years, and especially in the last few months, they'd been schooled in such esoterica as mineral identification, the characteristics of impact craters, the proper techniques for collecting samples. They had honed their skills on the lava flows in Hawaii and the desert of west Texas. When they left the earth they did so not only as pilots but as surrogates for the geologists who would have given their right arms to be in their place.

To save precious minutes on the surface, the geologists had sat down ahead of time and planned four possible traverses for Conrad and Bean to follow. About two weeks before the flight Conrad decided he wanted to carry maps, and the geologists had quickly drawn them up on a set of photographs. There wasn't time to get them on the official manifest, so Conrad arranged to have them stowed with his personal items. But that wasn't the only unofficial item Conrad had wanted to bring: a long-time collector of hats, he'd arranged for the crew systems people to make a giant, blue-and-white baseball cap that would fit over the top of his space helmet. He was going to put it on and wait for everybody on earth to notice as he bounded past the TV camera. Unfortunately, nobody could figure out a way to get it into the LM in secret. Never mind; Conrad had managed to smuggle

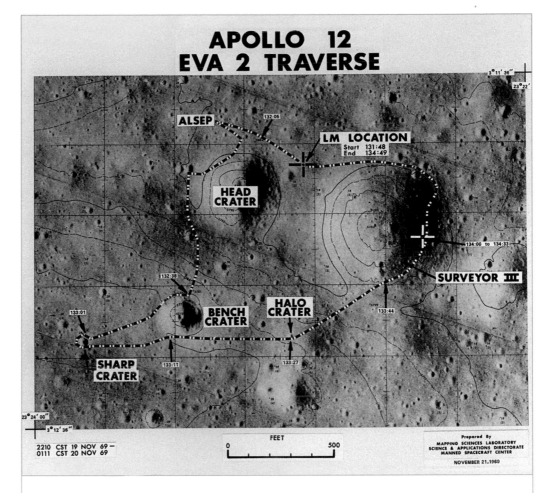

The head and belly of the Snowman are clearly visible in this NASA photomap of the path Conrad and Bean took during their second moonwalk. Beginning at the lunar module, they headed northeast to the lunar experiment site, then visited a succession of craters before visiting Surveyor 3.

something else in the pocket of his space suit, and when the time came, he and Bean would have their own little caper on the lunar surface.

It turned out that *Intrepid* lay right on traverse number 4, which followed a sort of misshapen circle around several craters. The geologists had explained before the mission that the moon's craters are like natural drill holes, that the ferocious energy of meteorite impacts had blasted chunks of rock from the crust and scattered them across the moon. Each crater of the Snowman was nothing less than a ready-made excavation into lunar history. By visiting different craters, the geologists hoped the moonwalkers would find out whether the lava flows that formed this region of the Ocean of Storms varied in age and composition. On earth, a geologist would have spent many days or even weeks on this kind of exploration, but the timeline gave Conrad and Bean a bit less than 2½ hours to get around the circle. Mindful of all they would try to accomplish, Conrad and Bean assembled a small assortment of gear—rock hammers, sample bags, core tubes, shovel, tongs, and maps—loaded them onto a portable tool carrier, and set off on a journey across the craters.

The first stop was the north rim of 360-foot Head crater, about 100 yards west of the LM. It didn't take long for the men to make a discovery. Bean looked at the places where Conrad's boots had dug into the gray soil and saw that they had uncovered a lighter gray, just underneath the surface. They could not hear the excited shouts of the geologists in the back room down the hall from mission control, but they knew they had found something significant. Before the mission, the geologists had told Conrad and Bean to look for evidence of a light-colored streak that was visible on the unmanned orbiter pictures—material that had probably been ejected from the impact that formed huge Copernicus crater, 230 miles to the north. Now it appeared they had found some. After the mission, this sample would let the geologists determine the age of the impact that had formed Copernicus, an important event in lunar history.

Conrad knew that he and Bean could easily have spent an hour at any one of these craters, but the geologists wanted as many different types of rocks as possible, and the clock was ticking. So he and Bean took off running, heading south to Bench crater.

Yesterday, during the first moonwalk, Bean had found that running on

Early in the second moonwalk, Bean runs past the steep-walled feature called Head crater. Beyond the crater's rim, just below the sun's glare, is the lunar module *Intrepid*, about 500 feet away. To the left is the tool carrier.

the moon was an experience all its own. He watched as Conrad skipped like a little boy and laughed with delight: "Wheee! Up one crater and over another. Does that look as good as it feels?"

Bean had discovered a better way. "Pete, bend and rock from side to side as you run. Like that. There you go." It wasn't really a run; it was more of a lope. Push off with one foot, shift your weight, and land on the other. Each step launched him into the air for long seconds, while he wondered whether he would land on a rock or in a pothole. To avoid this, he tried sticking his foot out to one side, but when he landed and pushed off again his uneven stance set him slowly rotating, giving him a new problem to deal with. He soon learned to anticipate each new step while he was still airborne, shifting his weight as he landed and immediately pushing off, setting up a rhythm, as if he were bounding across a rocky stream. It felt strange, and it was demanding—he couldn't take his eyes off the ground very often—but that only made it more fun. And the moon afforded him a luxury unknown to any runner on earth, a chance to relax in the midst of each step. It seemed, to Bean's amazement, that he would never get tired. ☾

"By the way," Conrad radioed Houston, "this is the smartest idea we've come up with. This map just works great out here." It worked great for Conrad, but most of the time Bean didn't have the foggiest idea where he was: from the surface, really big craters like Head and Bench looked nothing like the big, circular bowls on the map. Bean tried to navigate a few times on his own and gave up. It was beyond him how Conrad, running ahead of him, managed to study the map even as he bounded along. ☾

While Conrad led the way, Bean scanned the ground for something interesting. It wasn't easy to do field geology at a full gallop—and on the moon, it wasn't much easier standing still. Everything was so dust-covered that only the most subtle variations in texture were visible. Even the rocks looked almost exactly alike, until he held one right up to his faceplate. Then he could see hints of what lay beneath the dusty coating, the green glint of olivine crystals or a chunky white grain of feldspar. He could even see tiny pits, actually craters made by micrometeorites, peppering the sides of the rock, and he

Not far from Halo crater *(page 71),* **Bean hammers a core tube into the ground to sample the lunar soil to a depth of several feet. From these and similar specimens, scientists hoped to unravel the geological history of the moon.**

was telling Houston all about it when—"Hey, Al, quit baloneying and give me a hand"—Conrad cut him off. Bean understood; the geologists were going to have the rest of their lives to study these rocks; why waste time talking about them? And in the back of his mind Conrad was thinking, "Got to get around the circle." The spindly white probe—in Conrad's mind the mission's finish line—was waiting for them. Fortunately, Ed Gibson had good news: The moonwalk was being extended by half an hour.

"You know what I feel like, Al?"

"What?"

"Did you ever see those pictures of giraffes running in slow motion?"

Bean laughed; he knew what Conrad was talking about. He watched Conrad's feet as he ran along beside him. Each time his boots hit the surface they sprayed a small shower of dust, sailing out on perfect trajectories.

As they ran by the southern rim of Bench crater, they were beginning to feel the strain of their adventure. The entire run to the Surveyor crater was slightly uphill, and they could feel it. Their mouths were as dry as the desert from breathing pure oxygen. "Tell you one thing I'd go for," Conrad said, "is a good drink of ice water." ❝

To make matters more difficult for Bean, the tool carrier was getting heavy with rocks. He had to hold it up to his chest while he ran, to keep the rocks from bouncing out. This meant he was constantly fighting the arms of the suit. Before the mission he'd been so conscientious about exercise—every day he'd run a couple of miles on the hard beach at the Cape, pounding his legs into shape. And it turned out that most of the work was in the arms and especially the hands. He told Ed Gibson to pass on a message to Fred Haise, the lunar module pilot for Apollo 13, to do hand exercises.

Meanwhile, in Houston, the flight surgeon saw Conrad's and Bean's heart rates soar to 160 beats per minute. Ed Gibson passed on a message from the surgeon to the lunar surface, in code: "Pete and Al, we'd like an E-M-U check." EMU stood for Extravehicular Mobility Unit, in other words, the space suit and backpack. Ostensibly this was a request for each man to read out his oxygen supply and report any anomalies in suit pressure or the like, but in reality it let Gibson tell the men to cool it for a couple of minutes, without letting the whole world know about it. Conrad and Bean got the message. And they read out their oxygen quantity and the rest of it, but they couldn't bring themselves to stand around and do nothing. An interesting rock caught their eye, and they set about collecting it. Then they were on the run again. Everyone in mission control could hear them huffing and puffing.

The two men ran onward in their long, gliding, slow-motion-giraffe strides under the black dome of the lunar sky. Bean looked up as he ran, above the solar glare, and found a glittering crescent set in the velvet, exactly where it had been the day before. He looked down at his feet again, watching out for potholes and rocks, then leaned his head back for another glance. For just a moment, he silenced the voice in his head that always called him back to work, to the rule of the clock. And he talked to himself: *This is the moon. That is the earth. I'm really here.*

Suddenly, Bean felt his ears pop—his suit must be losing pressure! Heart pounding, he stopped and checked the gauge on his wrist. No change; it must have been some kind of weird transient. Only after the flight would the engineers decide that as Bean ran in the light gravity his body had bounced against the oxygen outflow port within his suit, closing it momentarily and causing a slight *increase* in suit pressure. But just now, that taste of fear was enough to put Bean's thoughts right back on the traverse. After that, he stopped looking up for a while.

THURSDAY, NOVEMBER 20
12:15 A.M.,
HOUSTON TIME

Standing on the rim of the Surveyor crater, Conrad and Bean rested for a moment. Just beyond the far wall they could see *Intrepid,* like a tiny replica. To their right, 300 feet away, sat Surveyor 3, gleaming in the morning sunlight. Antennas and sensors still reached upward from its tubular frame, just as they had on April 20, 1967, when the craft thumped onto the moon amid blasts from its braking rockets. Like *Intrepid,* it had the strange organic look of a craft designed for the void of space. Fuel tanks and batteries protruded from its open frame. This unlikely robot had made history when its mechanical claw, under command from earth, had scraped the skin of the moon and given scientists their first real information on the physical nature of the lunar soil. It had lasted just fifteen days, until the onset of lunar darkness. It was a relic now, but it was about to make one more contribution to space exploration.

There was some concern, both on the moon and in mission control, that Surveyor 3 might welcome its first visitors by sliding down the crater wall on top of them. For that reason Conrad and Bean had decided to approach from the side, following the contour of the sloping wall. This was a relatively old

Conrad pauses during the walk to Surveyor 3 to collect a rock sample near a small crater. He's using a long-handled scoop, selected from the tool carrier to his left, that lets him pick up a specimen without stooping or kneeling.

crater, and the geologists had warned them to expect a soft, thick dust blanket here, but the men were relieved to find firm ground that gave good footing. With slow, careful steps they closed in. They could see, to their surprise, that the Surveyor wasn't white any more, but a light tan. Had it been baked by the sun, or had Conrad and Bean sprayed it with dust when they landed? They'd investigate when they got closer, and they'd take pictures of it from every angle, and more pictures of the soil and rocks nearby, so the scientists could compare them with the Surveyor's own images from thirty-one months earlier. They'd collect some of those rocks, and then Conrad would snip off pieces of the craft to take home. But before they poked, prodded, and otherwise cannibalized the probe, there was something Al Bean and Pete Conrad wanted to do.

Before the mission, they'd had one of the support crew go out and buy an automatic timer for the Hasselblad, a little spring-loaded gadget. Conrad and Bean's idea was that they would mount the camera on the tool carrier and then pose, side by side, next to the Surveyor. It would only take a minute for them to fire off a few shots—saluting, waving, shaking hands, whatever— and Conrad was sure that when they got home one of those pictures would end up on the cover of *Life* magazine. He couldn't wait until everybody asked, Who took the picture?

Conrad had managed to smuggle the timer in the pocket of his space suit. He'd remembered to bring it into the LM with him, and just before they'd headed out on the traverse that morning, he'd dropped it into the tool carrier—which was now full of rocks and tenacious lunar dust. Bean rummaged in the bag for a moment, looking for a glint of chrome, but saw only grime. The only solution was to take all of the rocks out of the tool carrier. They couldn't talk about it; the whole world would know what they were up to. So they made hand signals. While Conrad held the tool carrier, Bean rummaged among the samples, each in their little Teflon baggies. He wondered if he should just put the tool carrier down, get on his knees, and lay the rocks on the ground, but he worried he'd never get them all back in the bag if he did. After a couple of minutes, Bean realized the timer was buried inside the bag, lost in the dust. He said quietly, "Forget it." ☾

12:38 A.M.

"Okay, Houston," said Conrad. "I'm jiggling it. The Surveyor is firmly planted here. That's no problem." Having assured mission control, Conrad wielded a pair of cutting shears. The engineers wanted some samples of metal tubing, to see how it had been affected by thirty-one months on the surface. But what Conrad wanted most of all was the Surveyor's TV camera. With all its circuitry and moving parts, it was the real prize. Conrad wrestled with the shears as he tried to cut through the camera's support struts. "Okay, two more tubes on that TV camera and that baby's ours." Another cut. "There's one," Conrad said. Now Bean gripped the camera with his gloves, and at last: "It's ours!" Conrad cackled. That TV camera was his trophy fish.

Back at *Intrepid,* Conrad was emptying the tool carrier into the rock box when suddenly out fell the Hasselblad timer. He called to Bean, "I've got something for you."

"Just what we need," Bean said. He picked up the timer and threw it into the distance as hard as he could.

Pete Conrad was exhausted. He sat on the floor of *Intrepid*'s tiny cabin, leaning against the wall in his grimy space suit. As tired as he was, he couldn't believe he and Bean had climbed into the LM after only 4 hours outside, with all that extra oxygen—maybe a couple of hours' worth—still in their backpacks. He wasn't about to tell Houston how frustrated he was; before the flight he'd agreed that if mission control gave him one time extension during each moonwalk, he wouldn't ask for a second one. So he and Bean had some time to kill. They ate lunch, and then they started into the checklist for liftoff. They were about an hour ahead of schedule, and when they came to the T minus 30 minutes mark in the count, they stopped, because the next steps —pressurizing the fuel tanks, for example—would have to wait until the proper time. For now, Conrad rested.

Conrad looked at Bean. He seemed nervous. There was absolutely nothing to do but watch the clock, but Bean was fidgeting with something on one of the instrument panels. What was he thinking? Conrad remembered how he had felt on his own first spaceflight, Gemini 5. Crammed into that tiny spacecraft with Gordo Cooper after a week in orbit, circling the earth over

Pete Conrad inspects Surveyor 3's TV camera, just before detaching it with metal cutters for return to earth.

and over, Conrad found himself thinking about the retro rockets, which had been soaking in the frigid cold of space for days on end, longer than anybody had ever been in space before. Conrad was terrified that when it came time to fire the retros nothing would happen—in which case he was sure he would slit his wrists. When the retros fired right on schedule, he breathed a huge sigh of relief. Now, on the moon, he was sure Bean was going through the same drill about the ascent engine. For his own part, Conrad had no such worries this time around.

Orbiting the moon, Bean thought,
was much more of a science fiction experience than
walking on it had been.

He'd already been through three launches in his career; this was just another one in a different place. In any case, if Conrad had any lingering doubts, he put them aside when he saw his lunar module pilot. Finally Conrad said, "Beano, are you worried about the engine?" Bean answered a quiet "Yep" and Conrad tried a bit of humor: "Well, there's no sense worrying about it, Al, because if it don't work, we're just gonna become the first permanent monument to the space program." Conrad wasn't sure the joke made any difference in Bean's outlook. But an hour later, when the moment of truth arrived, there were no unpleasant surprises. And then, as the Ocean of Storms fell away in one last, spectacular view, Pete Conrad and Al Bean headed for a reunion with a best friend in lunar orbit.

FRIDAY, NOVEMBER 21
ABOARD *YANKEE CLIPPER*,
IN LUNAR ORBIT

Bean was looking out the window. It was the first chance he'd had to relax and play tourist since arriving at the moon more than two days earlier. He thought about how strange it felt to orbit the moon. The strangest thing about it was the silence. There wasn't any engine noise, the way there always was in an airplane. And it felt odd to see the spacecraft fly in the same direction no matter how it was pointed. Orbiting the moon, Bean thought, was much more of a science-fiction experience than walking on it had been.

Flying in space was better than Bean had ever imagined during those long years as a rookie. Yesterday, during the rendezvous, he'd been slaving

During the trip home, Conrad, Gordon, and Bean became the first humans to witness an eclipse of the sun by the earth. The scene was recorded by the 16-mm movie camera inside the command module.

Dressed identically to meet the press shortly after Apollo 12's splashdown, Sue Bean *(left),* Barbara Gordon, and Jane Conrad mock the classic astronaut wife's response when asked about her feelings upon her husband's safe return from space.

away with the backup computer and the navigation charts while Conrad flew the lunar module. They had one more burn to do, and then they would have it made. And Conrad had said to him, "Why don't you just quit after this midcourse, and relax and enjoy it? You can take a minute and fly this vehicle." Startled by Pete's audacity, Bean wondered, wouldn't it put them off course? No, Conrad assured him, whatever digressions they made would be easy to correct. Bean was reluctant—surely mission control would know. Conrad laughed, "Not on the back side of the moon, they won't." Bean realized Conrad had planned this perfectly. And for a few minutes Bean had his hand at the crisp, responsive ascent stage. It was a moment that Bean would always remember as pure Pete Conrad, that in a small craft somewhere over the far side of the moon, he had taken the time to share with Bean a flying experience that even most astronauts would never know.

And then, Bean could see *Yankee Clipper* out ahead, growing slowly from a point of light into a gleaming spacecraft. He watched as the command module moved from sunlight into shadow and back again, thinking, *That looks so* neat. Dick Gordon did a perfect job on the docking—smooth as glass—and later, when he opened the tunnel to the LM, all he could see was two dim figures floating in a cloud of dust. He called down, "You guys ain't gonna mess up my nice clean spacecraft!" Before he would even let them into the command module, he made them take off their filthy suits. They passed them up through the tunnel and Gordon hurriedly zipped them into their stowage bags underneath the couches. Finally, when he and Conrad got up there, Gordon was happier than Bean had ever seen him. He was scurrying around, helping them stow their gear. He was offering them a drink of water. And in that moment, Bean was filled with a love for his crewmates that he had never experienced. Years later, Bean would say, his most special memories of the flight would not be about the moon or the earth; they would be about Pete Conrad and Dick Gordon.

Now Bean looked out at the bright, bleak cinder passing beyond his window. It was so utterly inhospitable. Everything in the universe has some function, Bean thought, but what is the function of the moon? Is it to make the tides? The earth would probably get along fine without them. Maybe it was as the geologists said: the moon is here to tell us the story that had been lost forever on our own planet. Maybe the moon would tell us where we came from. Bean didn't know the answer. As *Yankee Clipper* circled, Bean looked, and now and then, he wondered. He found himself thinking about the six-year journey that had gotten him here. He realized now, with his neck farther out than it had ever been, that life is too precious to spend it living by someone else's rules, even the unwritten ones of the Astronaut Office. He would be a good astronaut, but he would do it his way. As the moon bore silent witness, he told himself, "When I get back home—if I get back home—I'm going to live my life the way I want to."

ON THE WAY BACK TO EARTH

Pete Conrad wasn't feeling well. For one thing, he had somehow managed to come down with a cold. "How did I get the world's greatest cold on the moon, for Chrissakes?" Bean answered, "Because you had the world's greatest LMP with you on the moon, who had a cold." Where Bean's cold had come from, he had no idea. ❨

Then there was the rash on Conrad's chest—he'd had a reaction to something in the adhesive that stuck the biomedical sensors to his skin, and it raised itchy welts, like a bad case of poison ivy. It was almost the best moment of the flight when Houston finally told him he could take them off.

Bean was out of it, too—or at least it seemed that way to Conrad and Gordon. He was spending much of his time on the way back to earth sacked out in his sleeping bag. Not surprising—he'd peaked for the landing and the moonwalks, and the trip home was definitely an anticlimax. A couple of days ago Conrad would've been happy to fire up that SPS engine longer than the flight plan called for and come barrelling out of lunar orbit so fast that they would trim a day off the return. But Houston wasn't about to let them do that.

There was plenty of time to think on the way back, and Conrad did. He was proud of Apollo 12. They'd proven the pinpoint landing capability and accomplished their other objectives. Now astronauts could go to the places the scientists wanted them to. But along with the immense feelings of elation and satisfaction, there was something else, something Conrad hadn't expected. After seven years of eating, sleeping, breathing Apollo, he had finally had his mission, and it had been perfect, and now, just like that, it was all over.

A fragile chunk of basalt from the Ocean of Storms rests in a scientist's gloves at Houston's Lunar Receiving Laboratory. The rock had once been fractured, then coated with a thin layer of black glass, probably the result of a high-speed meteorite striking the moon.

The flying had been brief but challenging; the sights had been the most spectacular of his life. And yet, going to the moon wasn't what Conrad had expected. Yes, it was spectacular, but it wasn't . . . *momentous.* Looking back on it now, he couldn't shake the feeling that it had been so much like the training that it was almost an anticlimax. Take away the weightlessness and the view, and he might as well have been in the simulator. Seven years ago, he had told himself that if he made it to the moon he wouldn't let it change him; now, he had no worries that it would. Conrad kept these thoughts to himself as *Yankee Clipper* headed home, and he had no idea whether his crewmates felt the same way. He was quite surprised when Al Bean turned to him and said, as if he could read his mind, "It's kind of like the song: Is that all there is?" ☾

● ● ○ ○ ○ ○ ○ ● ●

Sometime after he and Dick Gordon and Al Bean got out of quarantine, Pete Conrad got his crew together for a survival meeting of sorts. It was time to talk about their futures. For his own part, Conrad had not known how he might feel about the moon after the flight, but now that he was home, he already knew he wanted to go back. It was a spectacular place, and he had proved to himself that it was also a fine workplace. Before the flight, he'd always daydreamed that he and Gordon and Bean would each return to the moon, as commanders of their own missions: Gordon would land on Apollo 18, Bean on 19, and himself on 20. There had been talk about bringing along a one-man lunar flyer for those last two missions, to expand the range of an astronaut in search of discoveries. In his daydreams he and Bean would develop the flyer and then test it out on their missions. But the flyer had been canceled in favor of a four-wheeled lunar roving vehicle. And it was clear now that budget cuts would force NASA to cancel the final lunar landing, Apollo 20, to free up a Saturn V for its earth-orbit space station project. Apollo 18 and 19 didn't look much more secure. For all their jokes about staying together as a crew forever, Conrad, Gordon, and Bean knew they would have to break up.

Conrad planned to get out of Apollo and onto the space station project, because that's where the available seats were. He told his crew they should do the same, and Bean planned to take his advice. But Dick Gordon decided to stay on. Going by the pattern of Deke Slayton's crew rotation, he would probably be assigned as the backup commander for Apollo 15, and then command Apollo 18. Conrad reminded him that Apollo 18 could disappear, but Gordon said he would take his chances. Having come so close to his goal, he couldn't give up now.

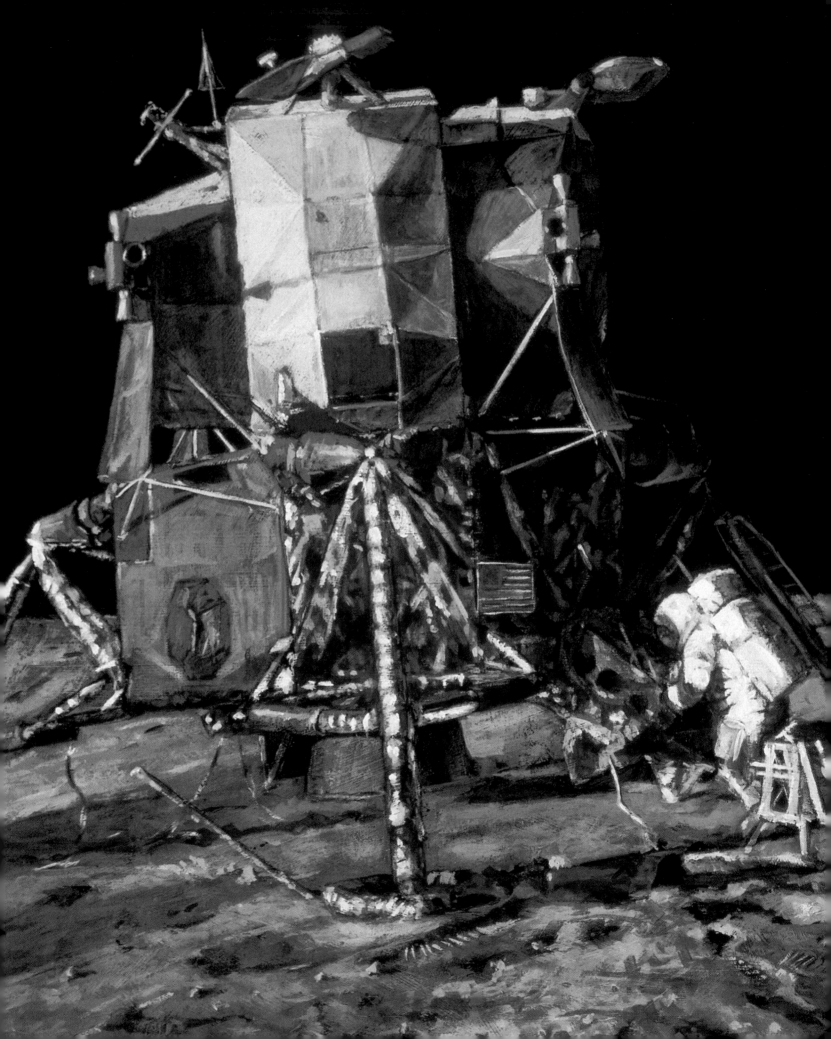

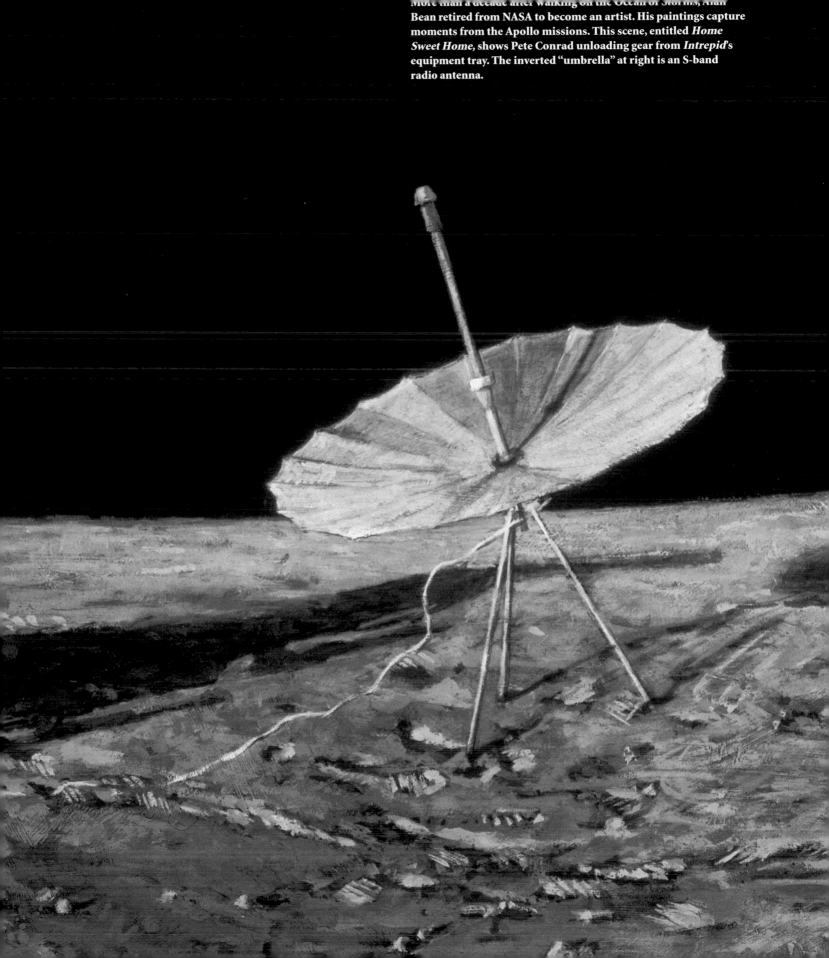

More than a decade after walking on the Ocean of Storms, Alan Bean retired from NASA to become an artist. His paintings capture moments from the Apollo missions. This scene, entitled *Home Sweet Home,* shows Pete Conrad unloading gear from *Intrepid*'s equipment tray. The inverted "umbrella" at right is an S-band radio antenna.

In Bean's painting *Helping Hands*, he and Conrad dismantle the core tube like the one seen on page 74. Working with Conrad, Bean says, provided some of his favorite memories of Apollo 12.

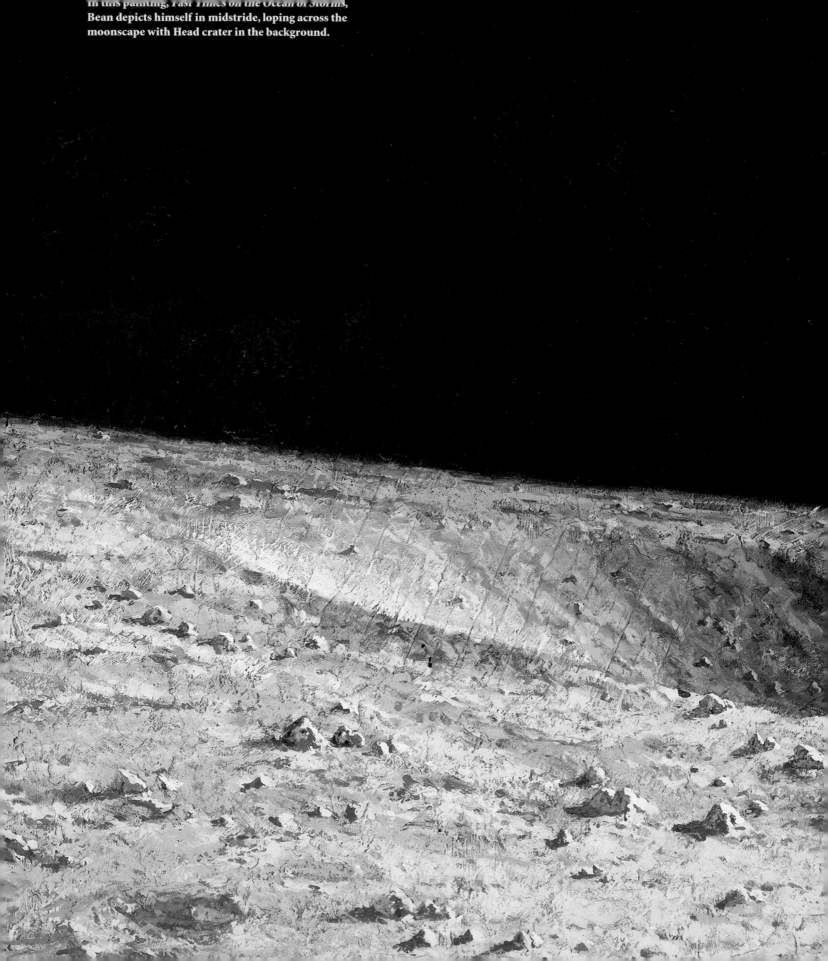

In this painting, *Fast Times on the Ocean of Storms,* Bean depicts himself in midstride, loping across the moonscape with Head crater in the background.

THE CROWN OF AN ASTRONAUT'S CAREER

APOLLO 13

I: A CHANGE OF FORTUNE

Jim Lovell carries four-year-old son Jeffrey on his shoulders as he looks ahead to his Apollo 13 mission. For Lovell, command of the third lunar landing marked the climax of a long and successful career as an astronaut.

The new decade opened with a nation polarized by the war in Vietnam and a space agency unsure of its future. After a year of triumph, NASA faced a combination of public apathy and outright hostility for costly, high-tech government "boondoggles" like putting men on the moon. At the same time, NASA's leaders were on uncertain footing with the Nixon White House. Already, faced with the leanest NASA budget in nine years, Tom Paine had suspended production of the Saturn V, leaving enough boosters to fly missions through Apollo 20. But Apollo Applications, the project to launch a temporary space station made from Apollo "spare parts," was already going ahead, and now it became clear that a Saturn V would be needed to launch the station into earth orbit. In January 1970, Paine canceled Apollo 20, and before long there were signs that two more Apollo flights were in jeopardy. ◖

Meanwhile, preparations went ahead for Apollo 13, set for a spring 1970 launch. But within NASA, the question was raised: Was it time to abandon the moon? In recent days none other than the MSC's director, Bob Gilruth, had privately called for an end to the moon landing program. ◖

A subdued Ken Mattingly looks over a flight plan in mission control shortly after Apollo 13's launch. Mattingly was still gloomy after doctors yanked him off the flight at the last minute, fearing that he might be infected by German measles.

Some called Gilruth the father of manned spaceflight. No one had done more to make Apollo a reality, and no one had higher regard for the astronauts. To them, too, he had always been something of a father. Even more than most of the safety-conscious NASA managers, Gilruth had sweated every new mission—but he had always been willing to take the risks, provided they were worthwhile. But there were new challenges and new programs on the horizon. With two lunar landings accomplished, was it wise to risk men's lives—and NASA's future—again and again? Gilruth wasn't at all sure that it was. Stop now, he said, before we lose somebody.

NASA was a democratic organization in which such opinions were freely expressed, but Gilruth's entreaties did not change the course of Apollo. Everyone knew how risky a venture it was, and sometimes it seemed as though they were tempting fate with each new mission. And no one realized this more than the astronauts. Any of them, if asked, would have said that it was just a matter of time before some hidden flaw in the system, some unnoticed mistake, that the quality control checks hadn't caught would come around to bite them. Spaceflight had always been a game of probability. But Jim Lovell, commander of Apollo 13, had no reason to suspect, as he and his crew headed moonward on an April evening, that the odds would finally turn against them that night.

MONDAY, APRIL 13, 1970
8:56 P.M., HOUSTON TIME

For the first time in weeks, Marilyn Lovell could finally relax. She was sitting in the VIP viewing room that looked out over mission control, watching a telecast from her husband's moonbound spacecraft, and what she saw cheered her. There was Jim, floating in the command module—the image was shadowy, and Jim's face was hard to make out at times—but to Marilyn it could not have been more heartening. Here was a portrait of a man in his element. Everything was going beautifully. Jim and his lunar module pilot, a rookie named Fred Haise, had just given a televised tour of their lander, *Aquarius,* and had displayed some of their gear—their space helmets, the hammocks they would sleep on, and special bags for drinking water that they would wear inside their space suits while they walked on the moon. Now, back in the command module *Odyssey,* Jim was acting as emcee.

"We might give you a quick shot of our entertainment aboard the space-craft," he said, holding up a small tape recorder. He brought it close to his communications microphone and turned it on; light piano-combo music filled the airwaves. Jim had no way of knowing that his words were going no further than the NASA centers; after two flawless lunar landing missions the networks weren't going to cut into the Doris Day show or the other prime-time programs to carry yet another telecast from space. Marilyn was here in the control center because it was the only place for her to watch.

"It's interesting," Jim said as he turned off the music, "to see that tape recorder floating there playing the theme from *2001: A Space Odyssey.* And of course, our tapes wouldn't be complete without 'Aquarius.' "

Marilyn was grateful things were going so well, grateful for the chance to feel good again. In the weeks before the flight she had felt an inexplicable sense of foreboding. Jim had made three flights before this one, including the first trip around the moon, but she had never been this anxious before any of those missions. It wasn't superstition, though she could confess to having had a twinge of that when she learned, the previous August, that Jim would command Apollo number 13. It was more a feeling that he had been lucky too long and that it was only a matter of time before his luck would run out. Still, she had not raised any objections, because she knew how much this flight meant to him. At last, he was going to do the thing he had wanted to do since he'd become an astronaut eight years before; he was going to land on the moon.

Last week, Marilyn had gone to the Cape to see Jim, and she had planned to return to Houston to watch the launch. But when it came time to leave, she found that she could not. So she stayed, to bid him farewell the way she had on Apollo 8: silently, while his rocket rose on a pillar of flame and disappeared into the sky. But as Jim had confided to her during a moonlit walk along the beach, Apollo 13 would be his last mission.

Also in the viewing room this Monday evening was a quiet young astronaut named Ken Mattingly. A member of the fifth astronaut group, he was thirty-

Gene Kranz (*foreground*) and his team of flight controllers watch a televised tour of the Apollo 13 lunar module on the evening of April 13, 1970. Minutes later, the serene pace of the mission was disrupted by a dire emergency in space.

four years old. Tall and thin, his brown, crew-cut hair almost gone, Mattingly was perhaps the most private man in the Astronaut Office. Even the other astronauts would not claim to know him well. But they would not have been surprised to learn that this evening Mattingly was in the depths of the worst depression of his life. Just a week ago, Mattingly had been the command module pilot on Apollo 13. Now he was a spectator.

The weekend before launch Charlie Duke, the backup lunar module pilot, came down with a case of German measles. He'd caught it from the child of a friend. The NASA doctors said Duke wasn't contagious and his illness had been incubating for two weeks, during which time he'd been to meetings and meals with Lovell, Mattingly, and Haise. Any one of them might also come down with German measles in the next two weeks. With less than a week until launch and the culmination of nine months of preparation, Lovell's crew were at the mercy of the doctors, who made daily blood tests and waited for the disease to show itself. By Wednesday the doctors had decided that Lovell and Haise were probably immune, but they could not be sure about Mattingly. His blood tests, they said, showed that he might already be fighting off the disease and could develop full-blown symptoms during the mission. By Thursday, Chuck Berry was recommending that Mattingly be pulled off the flight and that his backup, Jack Swigert, be sent in his place. And Mattingly was going through hell.

Like most pilots, Mattingly had an aversion to doctors. Pilots had a saying: There are only two ways you can walk out of a doctor's office—fine or grounded. More than one astronaut had sneaked off to a private physician, who was sworn to secrecy, rather than risk seeing a NASA doctor. And as the week wore on, Mattingly began to feel as though the doctors were not on his side. They were waking him up at 6 A.M. to draw blood. At 5 P.M. they drew more blood. Then they sent him to bed saying, "Now don't worry"—and then, the next morning, they'd wake him up again and say, "It doesn't look like you had a very good night's sleep." The whole process was enough to make anybody doubt his own sanity.

Aside from these hassles, Mattingly felt *absolutely fine.* But by Thursday, Swigert was being put through his paces in the command module simulator, and Deke Slayton was keeping a close eye on his work. Still, it was inconceivable to Mattingly that he would be left behind. He thought the doctors would either fill him up with so much gamma globulin he couldn't conceivably get sick, or else NASA would postpone the flight until the next launch window in May. His friends were consistently upbeat. If they believed his fate was sealed, they couldn't bring themselves to say so.

Friday, the day before launch, the decision was due. With Swigert getting all the simulator time, Mattingly had little to do, and he could only go running for so many miles on the space center's roadbeds. Slayton suggested he go flying. Mattingly drove to Patrick Air Force Base and did just that for a couple of hours, in a T-33. Afterward, he considered breaking the pre-mission medical quarantine to cheer himself up with some Dunkin' Donuts —those and the barbecue sandwiches at Fat Boy's were his main weaknesses at the Cape—but decided against it. He was wearing his blue NASA flight suit; surely he'd be recognized. Heading back to the space center, he turned on the car radio in time to hear a newscast: "NASA has announced that it will replace one of the three astronauts of Apollo 13, Thomas K. Mattingly . . ."

Suddenly it was real. Mattingly was furious to find out this way, but he wasn't about to blame anyone, especially not Deke. By the time he had reached the crew quarters he'd accepted the news. Jim and Fred were still over in the simulator, and everyone else reacted to Mattingly's presence with an awkward silence. Mattingly didn't have any more idea of what to say than

A day before liftoff, Lovell *(left)* and Fred Haise *(right)* pose for a last-minute portrait with their new crewmate, Jack Swigert. Never before had the roster for a U.S. space mission been changed so close to a launch.

they did. Slayton asked him what he wanted to do, and he answered that he needed to get away before the launch. Fine, Slayton said; he could grab an airplane and fly home that evening. After dinner he wished Lovell, Haise, and Jack Swigert good luck. When he arrived at Patrick, a T-33 was on the ramp, ready and waiting. Flying through the early-spring night, Mattingly was surprised to find the normally laconic air-traffic controllers calling him on the radio and chatting up a storm. He realized they were trying to make sure that he was awake. When he landed at Ellington, the ground crew was already waiting to take care of his airplane; he had only to get in his car and go home. And he realized that Deke Slayton must have quietly arranged for all of this; it was just the kind of thing Deke would do, and never acknowledge.

Now, watching the television transmission from Apollo 13, Mattingly was still in the depression that had descended on him four days ago. A distraction would have helped, but he had nothing whatsoever to do except watch the progress of the mission he should have been on.

8:59 P.M.

Jim Lovell ended the telecast with a farewell that had the ring of a 1940s radio announcer. "This is the crew of Apollo 13 wishing everybody there a nice evening"—all he needed was a closing theme song—"and we're just about ready to close out our inspection of *Aquarius* and get back for a pleasant evening in *Odyssey*. Good night." The TV monitors went blank, and Marilyn left, pleased that her husband was having a good mission. Mattingly, meanwhile, remained in the viewing room, listening to the voices of his former crewmates. In a few minutes, he and a friend planned to go out and have a beer.

●◑○○○○○◐●

At age forty-two, Jim Lovell was the most traveled man alive. With three spaceflights under his belt, he had racked up 572 hours and nearly 7 million miles, more than any other astronaut or cosmonaut. For many men that would have been enough, but not for Jim Lovell. From the day he joined the astronaut corps in 1962, his ultimate goal had been to command a lunar mission. Command was even more important to Jim Lovell than landing on the moon, and most veteran astronauts felt the same way. When Borman turned down Deke Slayton's tentative offer to fly the first lunar landing it had everything to do with the fact that he had been the commander on Apollo 8. If Lovell had any disappointment about his commander's decision, it vanished when Slayton assigned him to lead Neil Armstrong's backup crew. Within weeks after Armstrong's team came back from the moon, Lovell and his crew were training for their own landing.

By the spring of 1970, most of Lovell's colleagues from the second astronaut group had moved on to other things. Neil Armstrong had disappeared into the world of postflight P.R. that greeted him on his return from Apollo 11; it seemed unlikely he would fly again. Jim McDivitt had traded the demands of flying Apollo missions for the equally demanding job of preparing for them, as manager of the Apollo Spacecraft Program Office in Houston. Tom Stafford, though still on flight status, had replaced Al Shepard as chief of

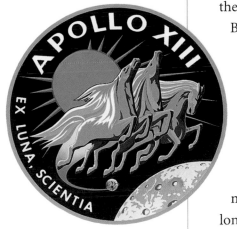

The mission emblem for Apollo 13, designed by artist Lumen Winter, featured three horses pulling Apollo's sun chariot from the earth to the moon. The Latin motto means "From the moon, knowledge."

the Astronaut Office, and wasn't scheduled for another mission. And Frank Borman had begun a new life as a vice president with Eastern Airlines. One day Borman came by for a visit while Lovell was in the simulator, and he seemed glad to be free of the training grind. "Jim," he said, "aren't you tired of this? I wouldn't want to go through this again."

Lovell couldn't have felt more differently. This was his seventh time around, counting the stints on backup crews, and even now his appetite for spaceflight was undiminished. He himself described it as an addiction. He could have gone on until NASA said he was too old to fly any more, but he knew that when he came back from Apollo 13 he would face a long wait, perhaps several years, before he flew again. Well aware of the astronauts still waiting for their first flights, he decided he would not get back on line for a fifth. Apollo 13 would be a great finale to a long spaceflight career.

Like every commander, Lovell wanted his mission to stand out, but he couldn't see why people would remember the third lunar landing. And that was fine with him. He wanted badly to land on the moon, and he was glad for the chance to make a contribution to science. The Apollo 13 mission patch read "Ex Luna, Scientia"—From the moon, knowledge—and Lovell thought of that when he christened his lunar module *Aquarius,* after the god of the ancient Egyptians who brought life to the Nile Valley (not to mention the popular song from the Broadway musical *Hair*). The command module *Odyssey,* meanwhile, took its name not only from Homer's epic work but from Arthur C. Clarke's science-fiction vision of space travel. When Lovell was back on earth, he would find irony in odyssey's dictionary definition: a long voyage with many changes of fortune.

The TV show was over, and Lovell waited in the lower equipment bay for Haise to finish closing up *Aquarius* for the night. They were about to shoot some pictures of Comet Bennett, an icy celestial wanderer that had graced the spring skies. But first mission control wanted Jack Swigert to stir up the service module's tanks of cryogenic liquid hydrogen and oxygen. In zero g, the super-cold fluids tended to become stratified, making it difficult for the astronauts and controllers in Houston to get accurate quantity readings. To remedy the problem, each tank contained a fan that acted like an egg beater to stir the contents. Strapped into the left-hand couch, Swigert flipped the switches marked H_2 FANS and O_2 FANS and waited several seconds, then turned the fans off. A moment later there was a loud, dull bang.

At first Lovell thought he knew what it was. The LM's cabin-repressurization valve made a bang whenever it was activated, and Haise had enjoyed scaring his crewmates half to death with it. Looking into the tunnel, Lovell called out, "Fred, do you know what that noise was?" Haise's dark eyes registered surprise; he hadn't touched the noisy pressure valve. Suddenly the master alarm rang in their headsets.

In *Odyssey*'s left seat, Jack Swigert looked up at the caution-and-warning panel and was alarmed to see that the light labeled MAIN BUS B UNDERVOLT glowed red. Main Bus B was one of the command module's electrical junctions, and the warning light could only mean that something had disrupted the flow of power to the command module's systems, perhaps an electrical short; if so, it was very bad news. Swigert called to Lovell, "The MAIN B light is on."

The electrical system was Fred Haise's specialty, and now Swigert called out to him to come back to the command module. In the meantime he floated over to Haise's side and scanned the electrical gauges. To his surprise, the voltage and current readings looked normal, and so did the fuel cells. Maybe there had been some kind of momentary glitch. But the bang—he'd *felt* it. The whole spacecraft had shuddered. It had scared the hell out of him. With only slight urgency in his voice, he called mission control. "Okay, Houston; we've had a problem."

"This is Houston," said Jack Lousma. "Say again, please."

Lovell was back in the center couch now. He keyed his mike. "Houston, we've had a problem. We've had a MAIN B BUS UNDERVOLT."

"Roger, MAIN B UNDERVOLT." There was a pause. "Okay, stand by, 13; we're looking at it."

Swigert could see no clear explanation on *Odyssey*'s instrument panel. Something must have happened to the lunar module. He wondered if a meteorite had struck it. The chances were supposed to be infinitesimal, but it would only take one sand-sized particle, moving at 8 or 9 miles per second, to cause crippling damage. If *Aquarius* had been hit, it could be losing pressure this very second. Swigert glanced toward the open hatchway and had the same thought that comes to men in a submarine taking on water: seal off the damage. As soon as Haise reappeared, Lovell and Swigert tried to install the forward hatch at the mouth of the tunnel, but it would not close; in their haste they had misaligned it. Suddenly they realized there was no air leak; apparently, nothing was wrong with *Aquarius* at all. But as Fred Haise was now discovering, something was very wrong with *Odyssey*.

When he heard the bang, and then Swigert's call, Haise had dropped what he was doing in the LM and headed for the command module. As he floated through the tunnel he heard pinging and popping noises: the sound of metal flexing. The walls of the tunnel were bending as the two joined spacecraft rocked back and forth against each other. Something had jolted Apollo 13. He heard the sound of *Odyssey*'s thrusters firing in response. Quickly, Haise made his way to his seat on the right side of the command module and began to investigate. *Odyssey*'s electric power came from three chemical-power plants called fuel cells, housed in the service module. Each fuel cell mixed hydrogen and oxygen to produce water; the by-product of this reaction was electricity. Current from the three cells flowed through two junctions, or buses, called A and B, which distributed power to run hundreds of different components. Everything in the spacecraft—the gauges on the instrument panel, the computer, the ball valves in the service module's SPS

With only slight urgency in his voice, he called mission control. "Okay, Houston; we've had a problem."

---

engine—ran on electricity. Haise checked the voltage on Bus B and the needle sank past the bottom of the scale; there wasn't even enough voltage to give a reading. As more warning lights came on Haise checked fuel cell number 3, the one that was supposed to be supplying power to Bus B, and saw that it was dead. The mission rules were printed on a card attached to the instrument panel, but Haise didn't even have to look at them. They were forbidden to go into lunar orbit unless all three fuel cells were working. Without being told, he knew the shattering disappointment of what had happened. They were not going to land on the moon. They had lost the mission.

But this was no time to dwell on that. Haise focused on *Odyssey*'s stricken electrical system. His moves came instinctively now, sharpened by hundreds of hours in the command module simulator. He began reconnecting *Odyssey*'s systems from the dead Bus B onto Bus A. Suddenly another red warning light flashed on—that bus too was starving for current. He checked fuel cell number 1; it was now dead too. Only one cell—number 2—was still producing current.

Lovell and Swigert, back in their couches, joined in the troubleshooting.

As the crisis aboard Apollo 13 deepens, Kranz stares intently at a monitor. When not serving at his flight director's post, Kranz spearheaded behind-the-scenes efforts to recover the astronauts.

The command ship and the attached lander had been set in motion by the bang and were still slowly turning. Swigert's best efforts didn't steady them, and he wondered whether some of the service module's maneuvering thrusters had stopped working. Meanwhile, Lovell checked the gauges for the oxygen tanks that fed the fuel cells and found strange readings: Just minutes ago they had both been full; now tank number 2 was completely empty, and tank 1's pressure had dropped to a third its normal reading and was still falling.

The dimensions of the problem grew with each passing moment. The three men had never seen so many seemingly unrelated malfunctions at once—maneuvering thrusters out, fuel cells dead, oxygen tanks losing pressure—even the computer had hiccupped during the bang. At their diabolical worst, the simulator instructors had never served up anything like this. There was an unwritten rule against simulating *multiple, unrelated, simultaneous* failures. In a spacecraft with thousands of separate components, the possible combinations of malfunctions would have been so numerous that it was impossible to train for them all. And everyone had always considered the odds of such a scenario very small.

In Houston, Gene Kranz was at the flight director's console, presiding over a team of flight controllers who could not believe what they were seeing. In the minutes since Swigert's first report of trouble, the situation aboard Apollo 13 had worsened on so many fronts that they suspected some kind of instrumentation failure had caused a slew of wrong readings. Even now, as they searched for some way out of the maze of problems, they wondered if they could believe their data. But that was because they were earthbound.

Lovell glanced out the side window next to Jack Swigert and what he saw gave him a queasy feeling in the pit of his stomach.

They had not heard the bang, or felt it the way Swigert had, or sensed its reverberations as the spacecraft slowly turned in response. And if they could have seen what Jim Lovell was seeing, right at that moment, they would never have blamed it on faulty instrumentation.

Lovell glanced out the side window next to Jack Swigert and what he saw gave him a queasy feeling in the pit of his stomach. A huge sheet of gas streamed from *Odyssey*'s side, swirling in the sunlight like cigarette smoke. For the first time he realized the depths of the crisis. He keyed his mike. "It looks to me that we are venting something. We're venting something out into space. It's a gas of some sort."

The gas was oxygen. When Swigert turned on the fans, he unknowingly triggered an electrical short inside a tank full of super-cold liquid oxygen. The bang that rocked Apollo 13 at 2 days, 7 hours, 54 minutes Mission Elapsed Time was the number 2 oxygen tank exploding. Every one of the failures that followed was a result. The jolt from the explosion caused the fuel cells' reactant valves to snap shut, cutting off their supply of oxygen and hydrogen and starving the electrical system. It also closed valves in the propellant lines that fed the maneuvering thrusters, making it very difficult for Swigert to steady the spacecraft, which was turning in response to the

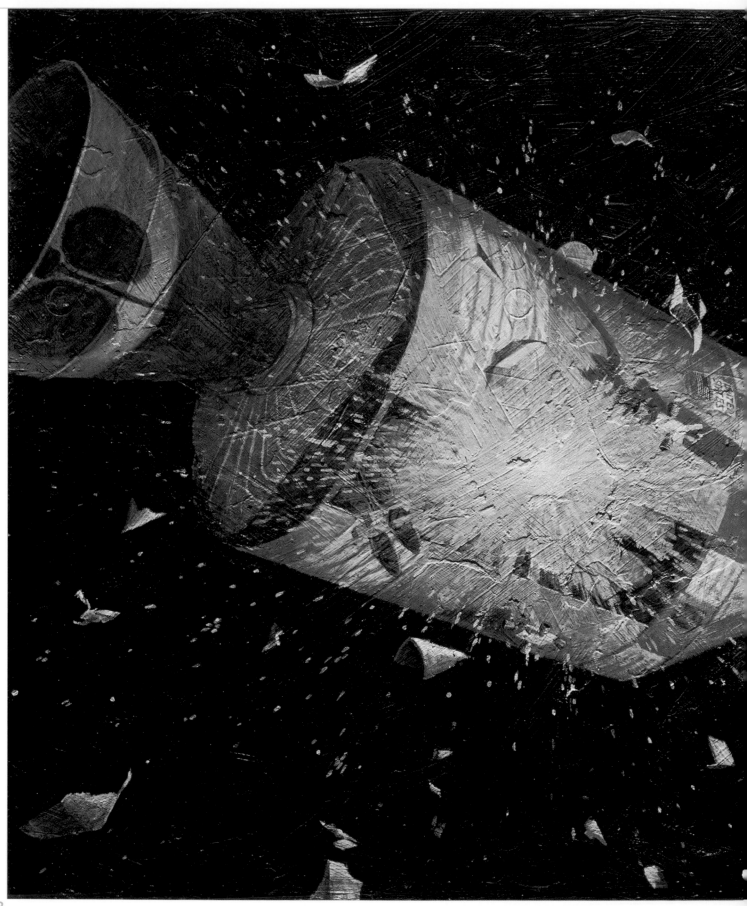

The explosion of Apollo 13's number 2 oxygen tank is portrayed by astronaut and artist Alan Bean in this painting entitled *Houston, We Have a Problem.* Debris streams away from the site of the blast, which took place within Apollo 13's service module.

propulsive action of the venting gas. It caused the computer to stop in the middle of its work and suddenly restart itself. And the blast tore out part of the plumbing for the service module's remaining oxygen tank, allowing its contents to spill overboard.

Lovell and his crew did not know what had happened inside their service module, and they could not lay eyes on it. The one thing they could see, the smoky haze spewing from its side, didn't single out either a micrometeorite hit or an internal explosion as the cause. But it wasn't important to know exactly why this was happening: all that mattered was that *Odyssey* was swiftly dying. It was only a matter of time before the last oxygen tank was empty and the last fuel cell was dead. At that point *Odyssey*'s only source of electricity would be the command module's batteries, but they had to be reserved for reentry. For Jim Lovell and his crew, the disappointment of losing the mission paled before the realization that they were in danger of losing their lives.

● ◐ ○ ○ ○ ○ ◑ ●

It took Marilyn Lovell about fifteen minutes to drive from the Mission Control Center to her home in Timber Cove, and when she arrived she phoned to invite her close friends and neighbors, Betty and Bob Benware, to come over for a drink. They declined; they had just come from one of the many Apollo 13 parties thrown by the aerospace contractors, and now they were tired. After Marilyn hung up, the doorbell rang; Pete and Jane Conrad had dropped by on Pete's brand-new red Honda motorcycle. They hadn't been there very long when a call came in for Pete on the "hot line," the direct link to the control center that NASA always installed in the crew's homes during a mission. As Pete went into another room to take the call, Marilyn's phone rang. It was another neighbor, Jerry Hammack, who ran the recovery operations for Apollo.

"Marilyn," Hammack said, "I just want you to know that countries from all around the world have offered to help with the recovery efforts, even the Soviets."

Marilyn couldn't make any sense out of what he was saying. She guessed he had been at the same party that had left the Benwares so tired. "Jerry," she asked, "have you been drinking?"

"Hasn't anyone told you?"

"Told me what?"

"The mission's been scrubbed."

She felt a wave of disappointment. "They're not going to land on the moon?"

IN 1970

Muhammad Ali refuses induction into the armed forces, ceding his world heavyweight crown to Joe Frazier.

Rhodesia declares itself a republic under the leadership of Ian Smith and severs ties with Britain.

The first Earth Day is celebrated.

Ohio National Guardsmen fire without warning on antiwar demonstrators at Kent State University, killing four students and wounding nine.

"There's no way they can, Marilyn."

Before she could find out any more, Marilyn looked up to see Pete Conrad at the top of the small staircase leading to the family room, motioning for her to hang up. His eyes were wide with alarm.

"Jerry," Marilyn said, "I'll have to talk to you later."

Pete came to her and quietly explained what he had just heard: "They've had a serious problem. They can't use the command module at all. . . . "

Suddenly the house was filled with her closest friends and neighbors. They stood in her living room in anguish. Someone turned on the television set and there was ABC's Jules Bergman, intoning the grim worst. He was all but saying the three astronauts were doomed. Marilyn felt the grip of panic. She hurried out of the living room, away from the crowd, and headed for the bedroom, and then into the master bathroom, locking the door behind her. In tears, she sank to her knees and prayed to God to bring Jim back to her alive.

"Thirteen, we've got lots and lots of people working on this. We'll get you some dope as soon as we have it, and you'll be the first to know." Jack Lousma was doing the best he could to be reassuring.

"Oh, thank you," Lovell said quickly. He noticed his mouth was dry.

Even as Lousma spoke, Lovell and his crew were trying to come to grips with their situation. They were 200,000 miles from earth, getting farther away every second. If only there had been enough electrical power left to operate the service module for even another hour, they might have been able to make what was called a direct abort, firing the SPS engine for all it was worth and executing a big U-turn in space. But even assuming the engine itself hadn't been damaged by the explosion—and there was no way to know—they could not operate it without electricity to open the valves to the combustion chamber, swivel the engine nozzle, and steer the spacecraft through the burn. The SPS was useless, and its 40,000 pounds of propellants were just dead weight. The bottom line was, they were going to the moon—or more precisely, they were going around it.

That would not have been a problem had Apollo 13 been on the free-return trajectory that Apollo 8, 10, and 11 had followed. But when Apollo crews started heading for landing sites in the western regions of the moon, it became necessary to change the timing of their arrival in order to ensure the proper lighting conditions for the landing. The solution was to make a slight

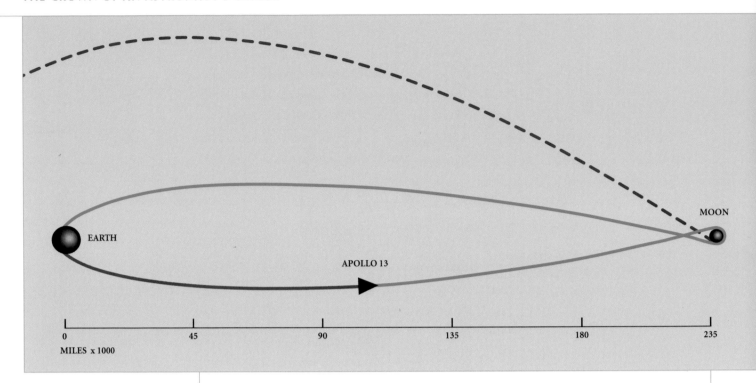

EARTH

APOLLO 13

MOON

| 0 | 45 | 90 | 135 | 180 | 235 |

MILES x 1000

As shown above, Apollo 13 *(triangle)* was nearly halfway to the moon when the oxygen tank exploded. Soon afterward, Lovell's crew had to fire the lunar module descent engine to place their spacecraft on the so-called free-return trajectory to earth *(green)*. Had they been unable to do so, they would have missed the earth by many thousands of miles *(red)*.

change in their approach to the moon. The new path, known as the hybrid trajectory, also had the advantage of saving fuel. The extra risk from giving up the free return was considered minimal. By firing the SPS to place them on the hybrid path, went the logic, the astronauts would demonstrate that the engine was healthy and would work properly to get them into and out of lunar orbit. The very act of throwing away their free ticket home would demonstrate that they didn't need it. As proof, Pete Conrad's crew had used the hybrid trajectory on Apollo 12 without mishap. It had always been assumed that any failure bad enough to cripple the SPS engine, with all its redundant parts—let alone one that could knock out an entire spacecraft— would kill the crew. No one had anticipated the situation that was shaping up this Monday night—three very live astronauts with a dead command module, heading for the moon.

Lovell and his crew knew that unless they did something to get back on the free return—and soon, before they went around the moon—they would miss the earth by some 45,000 miles. The only hope was at the other end of the tunnel: *Aquarius*. Its descent engine had far too little power for a direct abort, but it would suffice to get Apollo 13 back on the free return. Before they could fire the engine, however, they would first have to align *Aquarius*'s navigation platform with the stars, and one glance out the window told the men that would be all but impossible. The sky was filled with particles of debris from the explosion—in a cloud that observers on earth estimated to be 20 miles across—that caught the sunlight and shone like real stars. Furthermore, the LM's navigation telescope couldn't home in on stars automatically the way the command module's could. There was only one way to align the LM's platform: Lovell and Haise would have to copy the alignment from the command module's platform and feed the information into the lander's computer. And they would have to act fast: *Odyssey* was almost out of power. Already, Houston had instructed Swigert to turn off some of the systems; soon he would have to turn off the rest, including the platform and the computer.

Lovell and his crew had arrived at this basic plan when, after an hour and a half of troubleshooting, Jack Lousma radioed, "We're starting to think about the LM lifeboat."

"That's what we're thinking about, too," Swigert answered with an unwavering calm that typified the radio transmissions that had gone back and forth across the translunar gulf for the past 90 minutes. As Swigert spoke, the pressure in oxygen tank number 1 was slowly falling toward zero. Realizing that it was beyond saving, Haise left his seat in *Odyssey* and headed back through the tunnel to *Aquarius.*

●◐○○○○◑●

When the emergency began, Ken Mattingly had left the viewing room and taken a seat next to Jack Lousma, where he plugged in a headset. He listened to the voices on the flight director's loop as Gene Kranz and his controllers grappled with the bewildering mess that was unfolding 200,000 miles out in space. As an expert on the command module, Mattingly knew it was serious as soon as he heard Lovell and Swigert report the electrical failures. The crisis brought him right out of his funk. His thoughts now were of his crewmates, and how he might help bring them back.

It wasn't hard to understand why Mattingly, and everyone else in mission control, reacted to the emergency with disbelief. In this business there were certain well-based assumptions: Things broke, but they never did physical damage to the spacecraft in the process. It went without saying that oxygen tanks didn't explode. Listening to the voices of Kranz's men on the flight director's loop, Mattingly could tell that no one had a handle on the problem. The situation changed when Glynn Lunney took over the flight director's chair. A seasoned veteran at age thirty-three, Lunney had the uncanny ability to go right to the heart of the most complex problem. Now, as Mattingly listened, Lunney proceeded to give an inspiring performance. At first he, like Kranz, focused his efforts on trying to save the command module. When it became clear that this was impossible, Lunney put his LM experts to work coming up with a new checklist to let the astronauts power up *Aquarius* as quickly as possible. There were those in mission control, including one or two off-duty astronauts, who took one look at *Odyssey*'s oxygen readings and privately concluded that NASA had lost three good men. But Glynn Lunney didn't have time for any such thoughts. Mattingly heard him feeding rapid-fire questions and instructions to his controllers, keeping tabs on the dying command module, and checking the progress on the lunar module checklist. He and his men were now marching together toward a clear goal, to get Apollo 13 back on the free-return trajectory, and then home to earth.

Meanwhile, Gene Kranz organized controllers and specialists into teams to work on the rest of the recovery effort. How quickly should they bring Apollo 13 home? To answer that they would have to know how long *Aquarius*'s supplies would last. What about the command module? Could it be powered up after days in the cold of space? Unlike the crises that had threatened the first lunar landing and the launch of Apollo 12, this drama would play itself out over a span of days. Already the simulators were being cranked up so that astronauts could test new procedures devised by Kranz's teams. And across the country, thousands of NASA and contractor workers were mobilizing to save the three men in deep space.

11:13 P.M.

"Didn't think I'd be back so soon," Haise said as he set to work turning on the *Aquarius*'s systems. Time was critical now. As written, the shortest possible activation checklist took two hours, but there were only minutes of electrical power left in *Odyssey*. Jack Lousma passed up instructions from the LM experts, who were working furiously to pare the procedure down to the bare minimum. Meanwhile, inside *Odyssey*, where only the computer and the cabin lights were still turned on, Swigert had switched over to one of the command module's batteries, knowing that most of its power had to be saved for reentry.

While Haise worked, Lovell handled the critical transfer of the platform alignment from *Odyssey* to *Aquarius*. "Do it right," Haise cautioned his commander. "Take your time." To convert the data from *Odyssey*'s platform to *Aquarius*'s frame of reference, Lovell had to go through a fill-in-the-blanks conversion using a worksheet that was printed in the flight plan. He remembered that he had made errors when he'd done this during simulations, but

Two days after the accident, flight controllers confer with flight director Glynn Lunney *(seated, in jacket)* about a proposed recovery site in the South Pacific. At the helm of mission control during a crucial period early in the emergency, Lunney won praise for his masterful handling of the situation.

he and Haise both knew he could not afford a mistake now. "I want you to double-check my arithmetic on these numbers," Lovell radioed to Lousma. Once Lovell received confirmation from earth, Haise entered the data into *Aquarius*'s computer, and moments later the platform was aligned with the heavens. Minutes later, after Haise had activated the LM's maneuvering thrusters, there was no longer any need to keep *Odyssey* alive on precious battery power. Lovell yelled up the tunnel to Swigert: "Shut her down!" Nine minutes before midnight, Swigert turned off the last system, and *Odyssey* fell dark. Sometime after midnight, Lovell and Haise were in the midst of their work when Swigert floated into the LM cabin, a forlorn expression on his face. They understood; he had just put a friend to sleep. Swigert looked at his crewmates and said, "It's up to you now." ☾

Even now, oxygen spewed from *Odyssey*'s side like blood from a harpooned whale. The escaping gas acted like a small rocket, fighting Lovell's efforts to stabilize the joined craft—which the astronauts called "the stack"—with *Aquarius*'s thrusters. Lovell soon found that trying to control the stack from the lander was strange and awkward, like steering a loaded wheelbarrow down the street with a long broom handle. When he nudged the hand controller the joined craft wobbled unpredictably. It was, Lovell would say later, like learning to fly all over again. And he had to learn fast, because if he let the spacecraft drift uncontrolled, there was a danger that one of *Aquarius*'s gyros would be immobilized—a condition called *gimbal lock* that would ruin the alignment of the navigation platform. With no way of sighting on stars, there would be no hope of realigning it. Mindful of this, Haise said, "Why the hell are we maneuvering at all now?" The strain of the moment was in his voice. "Are we still venting?"

> *Better to burn up in the atmosphere, Lovell thought, than to become the first human beings never to return to their home planet.*

"I can't take that doggone roll out," Lovell said. Throughout the next 2 hours Lovell wrestled with his unwieldy craft, as the time for the free-return maneuver approached. He wondered if *Aquarius* would be able to point them toward home, and whether it would last long enough to get them there. Lovell and his crew had become the first astronauts to face the very real possibility of dying in space. If the free-return maneuver wasn't successful, Apollo 13 would miss the earth by thousands of miles, and continue to circle endlessly in the void. Years later, people would ask him whether he carried suicide pills, and he would answer that in all his years with NASA he had never heard of such a thing. There wasn't any need for them in a spacecraft; all that was necessary to take one's own life was to open the hatch. And even if he knew he were doomed, he would continue to transmit data back to earth as long as the radio held out, or until he and his crew had succumbed to lack of oxygen. But Lovell was determined to avoid that kind of death. At all costs, he told himself, he and his crew must get back to earth, even if they did not survive. Better to burn up in the atmosphere, Lovell thought, than to become the first human beings never to return to their home planet.

After an hour of preparation, Lovell and Haise were ready to fire *Aquarius*'s engine for the free-return maneuver. Apollo 13 was oriented correctly and held in position by the LM's computerized autopilot. Just let that computer keep working, Haise thought; he knew this burn required enough precision that orienting the spacecraft by hand might not be good enough. At one point during the preparations, Capcom Jack Lousma joked, "How do you like this sim?" Lovell answered wryly, "It's a beauty."

At 2:42 A.M. the descent engine rumbled to life at minimum power. Lovell waited 10 seconds, then slid the throttle to 40 percent. Lovell and Haise, standing, felt themselves pressed gently against the floor; Swigert settled onto the ascent engine cover. Twenty-one seconds later the computer shut down the engine automatically, precisely, perfectly. The burn was so good that Lovell and Haise didn't need to make any adjustments with the maneuvering thrusters. For the first time, the men let themselves believe they would make it back to earth. Another engine firing to hasten the homeward voyage still lay ahead, after Apollo 13 swung around the moon. But the worst was over—or so they thought.

II: THE MOON IS A HARSH MISTRESS

Fred Haise looked at his watch and was amazed to see that more than 6 hours had passed since the accident. He had been so absorbed in getting the LM activated, then preparing for the burn, that he had been completely unaware of time. Hours had passed this way, in intense activity that was punctuated by breathing spells as he waited for mission control to verify his work. During those brief respites, Haise registered a swift succession of images beyond *Aquarius*'s triangular windows: a cloud of debris particles sparkling in the black sky, the moon looming before him, big and close. He could scarcely believe how thoroughly his reality had changed in such a short time. It seemed

Few astronauts knew more about the intricate workings of the lander than Fred Haise, seen here in the lunar module simulator. His encyclopedic knowledge of the spacecraft became indispensable as he and his crewmates transformed the LM into a lifeboat for the trip back to earth.

like only minutes since he and Lovell were putting on their television show in this same tiny cabin.

Fred Haise, who had been flying for NASA since 1959, applied for the astronaut corps not out of any particular desire to go into space, but because it made sense to him to go where his agency was concentrating most of its resources. He was a member of the fifth astronaut group, the ones who called themselves "The Original 19." There was more than a trace of irony in the nickname. They were the third string, the benchwarmers, the "Red Shirts" of the Astronaut Office. Their chances of getting a seat on an Apollo mission seemed slim to nonexistent. But the Nineteen took the lead in one very important arena: hardware. With most of the more senior astronauts already in training for Gemini or Apollo, many of the Nineteen were assigned to represent the Astronaut Office at contractors' plants. Some, like Jack Swigert and Ken Mattingly, were detailed to Downey, to help shepherd the first command modules through testing. Others, including Haise and a navy pilot named Edgar Mitchell, were assigned to Grumman, and their life became the lunar module. Haise all but lived at the factory. He developed the kind of intimacy with the craft that only a "company pilot" gets. He knew far more about the guts of the lander than anyone needed to know in order to fly it. He knew what the relays looked like; he knew which pins on the electrical connectors went to which systems. He knew just about every one of the odd, one-of-a-kind parts that Grumman fashioned at the height of the weight-saving crunch. More nights than he could count, Haise had curled up on the floor of a lunar module and catnapped while a test was delayed. After two years of this he felt he knew the machine almost like a part of himself. He had unwavering confidence in *Aquarius,* but he also knew that he and his crewmates were asking it to perform in ways it had not been designed for. From here on, they were on untried ground. ☾

Aquarius was designed to support two men for 45 hours—a time span that included a systems checkout, the lunar landing, a 33-hour stay on the surface with two moonwalks, and the rendezvous with the command module. Now, Lovell's crew would ask it to support three men for perhaps twice as long—the journey back to earth would take anywhere from 77 to 100 hours, depending on what kind of engine firing the men made after rounding the moon. The situation was pure black and white: *Aquarius* would have to function at less than half its normal ration of supplies, or Lovell, Swigert, and Haise would not be alive when they reached earth. Immediately after he and Lovell returned to the LM, Haise began doing mental calculations of oxygen, power, water, and other consumables. Later, when there was a lull, he got out his books, where he had data on every one of the lander's systems, along with actual flight performance from previous lunar modules. Would *Aquarius*'s reserves be enough? Haise didn't even bother calculating oxygen. He knew without checking that there was more than enough for the trip home. First there were the oxygen tanks in the lander's descent and ascent stages. Then, the oxygen in the backpacks he and Lovell would have used on the moon. And finally, a set of emergency bottles in the command module. In all, there was enough oxygen to keep three men alive for 8 or 10 days—in other words, more than twice as much as they needed.

Haise's worry was electrical power. Unlike the command module, the LM

Ken Mattingly (*standing, at left*) looks on as some of his fellow astronauts review data from the crippled Apollo 13 spacecraft. From left: Deke Slayton, rookies Vance Brand and Jack Lousma, and John Young.

had no fuel cells and relied solely on batteries. Fully charged, *Aquarius*'s batteries were good for roughly 2 days of normal operation. It was obvious that the only way to stretch them to 4 days was to turn off most of the LM's systems—cabin lights, gauges, and after the next rocket firing was over, even the computer—and he knew mission control would have them do just that. Haise went down through the list of the items he knew they *wouldn't* turn off: radio, 1.29 amps; environmental control system, 5.85 amps. . . . Haise figured it would take about 18 or 20 amps to keep *Aquarius* going at a bare minimum —about as much as a 1970-vintage refrigerator and color television would consume together, and less than half the 50 amps the LM used in normal operation. *Aquarius*'s batteries could supply 20 amps for 100 hours —enough to see them home, and at best, perhaps a full day's worth more than they'd need. ❮

But battery power alone would not keep the LM running. Electronic gear gets hot, and it must be cooled to keep functioning. The most precious substance aboard *Aquarius* wasn't oxygen or food; it was cooling water, and Haise found that the 338 pounds stored in *Aquarius*'s tanks weren't enough to last the trip. By his calculations, *Aquarius* was probably going to run out of water about 5 hours before reentry. That would have been an ominous prediction if not for an important bit of data from Apollo 11. Before Neil Armstrong and his crew cast off *Eagle*'s ascent stage in lunar orbit, they began an experiment: they left everything inside the lander operating, but deliberately turned off the water supply. *Eagle* was a good soldier; it transmitted valuable data right up until it overheated and died. If *Aquarius* was as good a ship as *Eagle*, and Haise was sure that it was, then even if the water ran out, *Aquarius*'s systems would keep running for about 8 hours more. That left 3 hours to spare. When Haise had finished his calculations, he knew that if nothing else went wrong, they would probably make it. And when mission control called with their own assessment of the situation, Haise realized that the LM experts, more than 200,000 miles away, had reached the same conclusions.

With the free-return maneuver out of the way, Haise found himself exhausted and emotionally drained. The normal flight plan had gone out the window after the emergency, and with it, normal rest periods. It had been more than a full day since the men had slept, and Houston recommended they set up watches. Lovell decided that he and Swigert would stand watch in the LM while Haise slept in the command module; that way, either himself or Haise would always be available to take care of *Aquarius*. "I'll wake you up," Lovell said, and Haise headed for *Odyssey*.

6:24 A.M.

Jack Swigert and Jim Lovell floated in *Aquarius* while Haise slept in *Odyssey*. Every hour, Lovell had to use *Aquarius*'s thrusters to rotate the stack 90 degrees, so that the sun's warmth would be evenly distributed on the hull. Normally, this was done by the command module computer, but the LM's computer had no software to perform this slow-motion-barbecue spin; Lovell had to perform the maneuvers by hand. Even this simple task was made difficult by the unwieldy craft, but at length he managed it. Now, as the stack turned, the moon slid into view. Already it was close enough to see craters with the unaided eye.

"That old moon's getting bigger and bigger, Jack."

Jim Lovell and Jack Swigert did not know each other very well. They had arrived at NASA four years apart, and as a backup crewman, Swigert had not worked closely with Lovell until the day he took Ken Mattingly's place. In some ways Swigert and Mattingly were opposites. Both were bachelors, but that was where the similarity ended. Mattingly's whole life seemed to revolve around the program. He appeared to have no social life whatsoever; he was an ascetic monk at the monastery of Apollo. Once, when a reporter asked him whether he had any plans to get married, Mattingly blushed and jingled the change in his pocket. ❆

Swigert, meanwhile, was a world-class bachelor who pursued his calling

---

When it came to the emergency procedures for command module malfunctions, Swigert literally wrote the book.

---

with the same methodical, fastidious approach he brought to test flying. His date book included codes that signified single, divorced, or widowed. His success with women was at once legendary and utterly mysterious. More than one astronaut who teased Swigert about the way he dressed—he wore white socks with a suit and tie; his feet were flat and the sides of his shoes wore out before the soles—had to will his jaw shut when Swigert showed up with a drop-dead-gorgeous date. They knew Swigert was doing just as well wherever his T-38 took him. But those who worked with Swigert knew his dedication to Apollo. After the Fire, when it was time to get down to rewriting checklists and procedures, Swigert had been tireless. His battle cry was persistence: Even after NASA turned him down twice, for the second and third astronaut selections, Swigert was back to do battle with the ink blots and the treadmill in 1966; this time he won.

When talk about replacing Mattingly started, Lovell spoke to Tom Paine to try and stop it. He had nothing against Jack Swigert, but Mattingly knew the command module down to the last detail, and Lovell had come to depend on him. He told the NASA administrator that if Mattingly got sick it would probably be on the way home, with no impact on the mission. But Paine refused to overturn Chuck Berry's recommendation. It wasn't worth the risk,

he said, especially since an illness would bring NASA a great deal of negative publicity. Lovell realized he had no choice but to accept the situation.

But taking Swigert onto the crew was no simple matter. It was true that Swigert was a command module expert, having lived with the craft at North American the way Haise had done with the LM at Grumman. When it came to the emergency procedures for command module malfunctions, Swigert literally wrote the book. He was one of the only astronauts who actually went to Deke Slayton and asked to be a command module pilot.

The problem was that Swigert had been training with John Young and Charlie Duke for eight months. He'd forged a close working relationship with them; he was used to their techniques. Now, with launch only days away, he was switching crews. Would he be able to mesh with Lovell and Haise in the critical phases of the mission? During an emergency rendezvous there would be no chance to explain verbal shorthand like, "You've got the next burn." Lovell told Tom Paine he'd agree to the switch on one condition: This shotgun marriage would have to be tested in the simulator.

And so in the last days of their training, just when they should have been taking it a little easier, Lovell and Haise were in the simulator for hours on end, running through practice launches and rendezvous with Swigert. By Friday, they knew the marriage was going to work. But by launch day, the sadness they felt for Ken Mattingly and the annoyance with the daily blood tests and other medical hassles of the past week all but robbed them of the excitement of going. When the three men arrived at the White Room, sealed in their space suits, the usual exuberance of the closeout technicians was gone; they were subdued and nervous. Still, Lovell knew, if he and Haise had been through ups and downs, that was nothing compared to what Swigert had been through.

Any astronaut who had ever served on a backup crew knew that the experience is an emotional roller coaster. All through the long months of training, the backup man keeps himself motivated with the hope that he might get to go instead of the prime crewman. About a month before the flight, he realizes he's not going after all. A week before the launch of Apollo 13, Jack Swigert was in the same position as every backup crew member before him: he had been mentally and emotionally backed down from the mission for weeks. His biggest worry was finding hotel rooms on the beach for the prime crew's launch guests. Then Charlie Duke got German measles.

And as it turned out, Swigert had fit in extremely well, Lovell thought. In the first two days of the flight, while everything was still going normally, Swigert had proved the benefits of standardized training. He handled his role

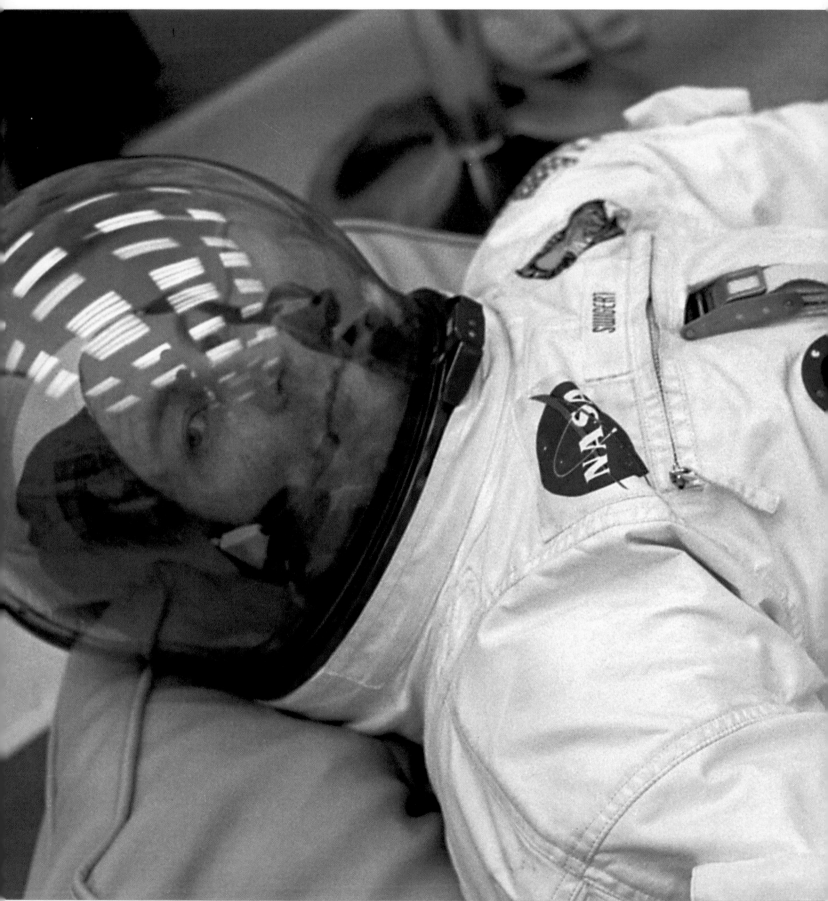

with professionalism and skill. As much as Lovell hated to lose Mattingly, he had to admit that things were probably going to work out okay—until Monday night. Lovell could only imagine what Swigert was thinking now. ☾

Since midnight Tuesday, Jack Swigert had been a man without a spacecraft. As long as they were in *Aquarius,* Swigert would not do any flying; he had barely set foot inside the LM simulator. He was a spectator, perched atop the can-shaped ascent engine cover in the back of the cabin, while Lovell and Haise worked. The lander seemed so unsettlingly delicate. At one point he was about to take a picture of the moon and Haise cautioned him not to hit the window with the camera; it might break! And he heard strange, distressing sounds. The LM whined and squealed; it gurgled. He asked Lovell and Haise, "Is there something wrong with the environmental control system?" They reassured him: those sounds are normal; the time to worry is when you *don't* hear them.

Now, Swigert watched as Lovell struggled to turn the spacecraft. Through the windows, he could see the moon as a pockmarked half-circle; then the earth, tiny and radiant. It had looked small before the bang; now it seemed even smaller. It was hard for Swigert to accept the fact that the best way to get back there was to keep going away from it. About 12 hours from now, at 6:30 this evening, Apollo 13 would swing around the moon and make its closest approach, or pericynthion, above the far side. Two hours later, according to the plan devised by mission control, the men would fire up *Aquarius's* descent engine once more. Much would depend on that firing, which was known as the PC + 2 burn, short for "pericynthion plus 2 hours." Without it, the trip home would take a total of 4 days; with it, that time could be trimmed to as little as 2½ days. The increased margins on *Aquarius's* supplies could mean the difference between making it and not making it. In addition, it would shift the splashdown point from the Indian Ocean, where the navy had few ships, to the South Pacific, where the recovery force was now stationed. But Swigert, with less confidence in this unfamiliar machine than in his own, looked ahead to the PC + 2 burn with some anxiety. He asked Lovell, "Do they think we'll have any trouble doing the next burn . . . ?"

"Probably not, if everything holds together," was Lovell's assessment.

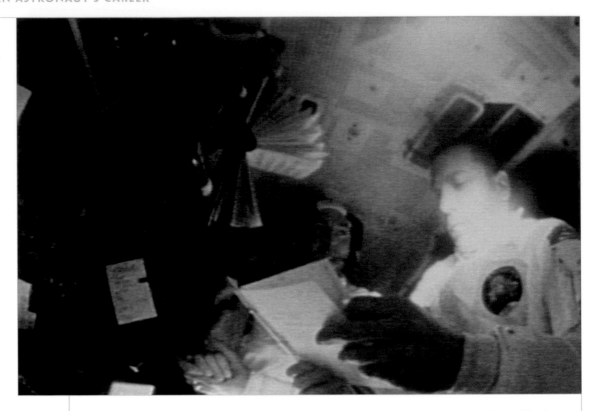

Swigert (*right*) reviews a checklist while Lovell rubs his hands against the chill in the tiny cabin of *Aquarius.* Fred Haise captured the scene with the onboard 16-mm movie camera.

"We've already made one burn." But then, in a matter-of-fact tone, he added, "Well, Jack, this is going to be touch and go." He seemed to be talking about the entire journey home, not just the next burn.

TUESDAY MORNING

TIMBER COVE, TEXAS

For Marilyn Lovell, the past twelve hours had been a vigil by the squawk box. The men had turned off their radio's amplifier to save power, and sometimes their words were nearly drowned out by static. It helped to have some of the other astronauts in the house, including Pete Conrad, explain to her what was going on. And she took some measure of comfort from the presence of friends, including Susan Borman, Jane Conrad, and Betty Benware. Marilyn's memory of these next hours would be hazy; she would be told later that she lit one cigarette after another, setting them down here and there while friends followed behind putting them out.

In the morning, she sent young Jeffrey, age four, off to nursery school as if nothing were wrong, but her two daughters stayed home. She had already talked by phone to her son James, fifteen, who was at a Wisconsin military academy. Earlier, Marilyn had worried about Jim's mother, who had recently suffered a stroke and was in a nearby nursing home; if she found out about the emergency, Marilyn feared it might kill her. So Marilyn called the home

In Timber Cove, Texas, Marilyn Lovell *(below)* kept an exhausted vigil by the "squawk box" relaying communications from Apollo 13, as did Mary Haise *(bottom)*—seven months pregnant—in Nassau Bay.

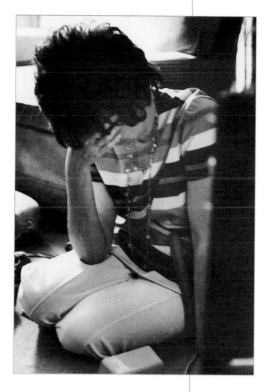

to have the television in her room turned off. But in the Lovell house, the TV was on, and ABC News's gloomy prognosis so upset Marilyn's younger daughter, eleven-year-old Susan, that the girl burst into tears. "He's coming back," Marilyn reassured her. "There's just no reason why he isn't coming back." Marilyn knew it was a white lie. She'd asked some of the NASA people about the odds of saving the men, and they'd been honest with her: 10 percent. But she also knew that she could not give in to her terror, imagining awful scenarios the way she had when Jim was a test pilot at Patuxent River. ❮

Besides, other people had already done that for her. Last year, she and Jim had gone to the premier of a movie called *Marooned* in Houston. In the movie, three Apollo astronauts are stranded in earth orbit after their engine fails to ignite for retrofire. Before long, mission control has run out of ideas, and the head of NASA's manned space program, played by Gregory Peck, is spelling out the doom scenario to a closed-door meeting of his deputies: "If necessary, on Wednesday morning, the President will issue a message to the nation emphasizing the courage and determination of the crew, and their final wish—that the program be continued, without pause." But the chief astronaut, played by David Janssen, will have none of it. He slams the table and barks, "I don't give a damn about the next mission!" Before long, he's convinced NASA to launch a rescue mission with himself as the pilot. And the rescue succeeds— but not before the mission commander, whose name is Jim, is killed in a space-walking attempt to fix the engine. Marilyn did not find *Marooned* very entertaining.

But the movie was far from Marilyn's thoughts. She knew she must put fear out of her mind, or she would not last through this ordeal. After Father Raish came by to conduct a private Mass, her spirits were lifted, and she carried herself with a calm that amazed her friends. Susan Borman said, "I hope the navy's proud of you."

Meanwhile, the squawk box relayed the voices of Jim and his crew, amid static. The men were on hot-mike, and every once in a while it was possible to hear them talking to one another. Their voices did not show any signs of strain. She knew Jim's attitude about panic: Where would it get him? After you bounce off the walls a few times, Jim would say, you're right back where you started.

Right now, Jim was telling Jack Swigert to fill up some juice bags

A day after the explosion, Father Donald Raish leads a private Mass in the Lovell home for Marilyn *(striped dress)*, other wives, and friends. On the wall are mementos from Jim's earlier Gemini and Apollo flights.

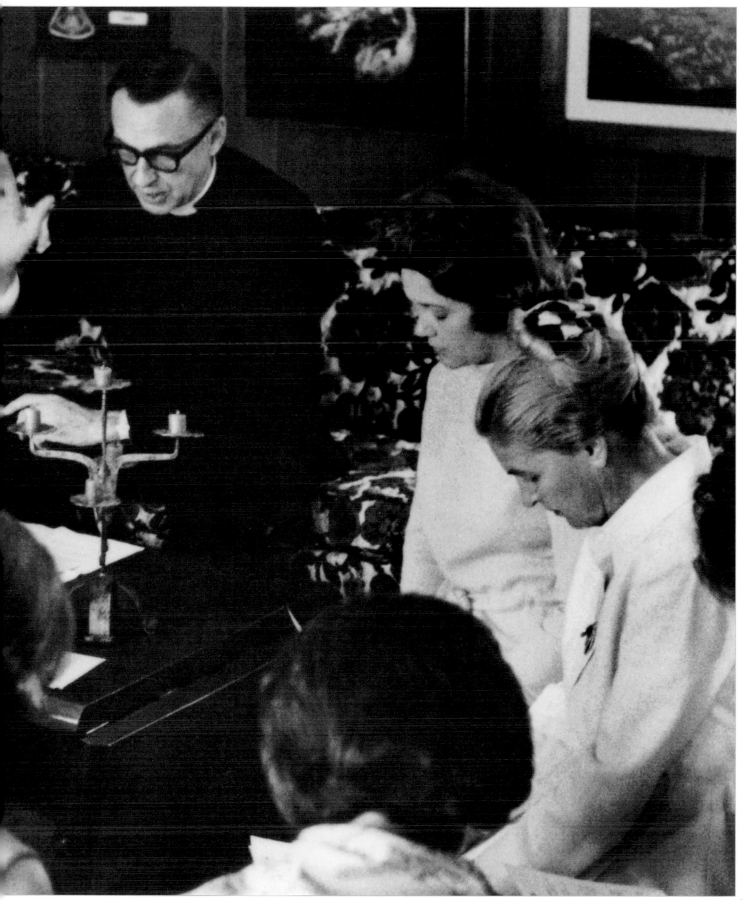

with drinking water from the command module. And then, it was possible to hear Jack greet a returning Fred Haise: "Come on in, Fred-o. Did you sleep good?" Not long after, Jim Lovell could be heard saying, with some fatigue in his voice, "Lookit, I gotta get some sleep."

12:42 P.M.

Jim Lovell was back in *Aquarius* after only 3 hours. Unable to stop thinking about the situation, he hadn't slept. Had they missed anything? Could they find a way to stretch the LM's supplies even further? He'd seen Haise's numbers, but he had a hard time believing them. He wondered about the forecasts coming from mission control—they were a little too positive, a little too vague, and it occurred to Lovell that things might be worse than they were willing to admit. Lovell's thoughts turned to the PC + 2 burn, set for about 7½ hours from now. At that point, some 20 hours would have elapsed since he and Haise had aligned *Aquarius*'s platform. Lovell knew that there was a chance the gyros could have drifted out of alignment, perhaps only slightly, but still enough to affect the accuracy of the maneuver. There was no way to check it against the stars; the sky beyond *Aquarius*'s windows was still a blizzard of debris particles. The only chance to see stars would be during the half-hour that Apollo 13 flew through the moon's shadow. Lovell knew he might eat up those precious minutes simply maneuvering the stack to bring stars into sight of the LM's telescope. There had to be some other way. ☾

Houston had figured out an answer, and at 1:42 P.M., a now-healthy Charlie Duke took the Capcom mike to read up the procedure, and Lovell didn't like what he heard. He and Haise were to test the alignment by sighting on the one star they could see, the sun. Mission control would send up a set of coordinates, which Lovell would use to steer the craft into position. If the alignment was good, the sun would show up in the eyepiece of the navigation telescope. It didn't have to be perfect, they said; even a 1 degree error in alignment would be okay—as long as the telescope crosshairs were touching the sun's disk. Lovell agreed, but he had real doubts it would work. But the alternative—sighting on stars during the brief minutes in the lunar shadow—he liked even less.

It was around 3:00 P.M. when Lovell wrestled Apollo 13 into position. The task was even more difficult now, because the attitude indicator, or 8-ball, had been turned off to save power, and judging orientation from the computer's readouts was awkward. In Houston, flight director Gerry

Griffin waited with great anxiety; he knew this was one of the most critical moments of the flight. Lovell wished Griffin and his controllers could look over his shoulder, but they told him now that they were having trouble locking up on *Aquarius*'s telemetry. Lovell would have to do the best he could without them.

Now the solar glare filled *Aquarius*'s windows. While Lovell sighted through the telescope, Haise checked the readouts. Lovell made a correction to the attitude, then moved aside and let Haise look. Lovell asked anxiously, "What have you got?"

Through the dark protective filter, Haise saw a bright arc. "Upper right corner of the sun."

"We've *got* it," Lovell exclaimed. In Houston, the flight controllers cheered, and Gerry Griffin's hands trembled as he wrote the successful result in his log.

Nearly three hours later, Apollo 13 flew into the moon's shadow. Outside, where the sky was suddenly filled with stars, the men could see dark silhouettes which they realized were debris from their explosion. Somewhere out ahead was the moon. Lovell knew he would not see it nearly as well this time as he had on Apollo 8, for Apollo 13 would come no closer than 136 miles. The encounter would be a fleeting one, for they would be moving so fast it would take just half an hour to round the far side, and only 20 minutes of that would be in sunlight. For Jim Lovell, the small, lifeless world that had been his goal for eight years was now simply a way point along a troubled journey. At 6:21 P.M. his headset filled with static as Apollo 13 flew out of contact with earth.

6:29 P.M.

With blinding suddenness the sun reappeared—reassuringly, it was exactly when Houston had predicted—and up in *Odyssey* Jack Swigert and Fred Haise were glued to the windows. Nothing could have prepared them for a sight as alien as the far side of the moon, and for the first time since the accident they forgot their life-or-death situation. As the craft sped past the cratered ball, Swigert and Haise fired off pictures.

"If we don't get this burn off," Lovell told them impatiently, "you won't get your pictures developed."

"Relax, Jim," Haise said. "You've been here before, and we haven't." Like a

The astronauts had this view through *Aquarius*'s overhead window as their docked spacecraft rounded the moon. At right is the curved hull of the command module *Odyssey*.

Heading toward re-entry, Jack Swigert turns to the check-list to reactivate the command module for reentry. Mean-while on earth, controllers worked feverishly to update the checklist with new instructions for the power-up, which they radioed to the crew.

billiard ball in bank shot, Apollo 13 rounded the moon, its path bent by the lunar gravity. Soon it would be pointing back toward earth. And now Lovell joined his crew at the windows, pointing out places that only he had seen be-fore, until Houston was on the radio once more. As the full moon shrank into the distance, Lovell and Haise went back to work.

8:40 P.M.

Because of Lovell's anxiety to get ready for the PC + 2 burn, he and Haise fin-ished powering up the LM almost an hour ahead of schedule. Lovell thought of all the power they were wasting just sitting there waiting, and he chided himself for jumping the gun. Finally, he and Haise stood at *Aquarius*'s con-trols, ready for the maneuver. Because the command module was attached, a situation the LM's software wasn't written to accommodate, Lovell had to fire the engine by hand. When the time came he pushed the ENGINE START button, waited 5 seconds, and slid the throttle to 40 percent. "We're burning," he told Houston. Twenty-one seconds later, the computer throttled up to full power. Lovell was amazed at how smooth and quiet it was; this engine was a beauty. He wished he were using it to land on the moon, right then.

●●○○○○○●●

For the first time since the accident, Jack Swigert's spirits lifted. Every second brought them closer to home instead of farther away from it. If *Aquarius* held out for another 62 hours—and based on mission control's latest estimates, there seemed no doubt that it would—they would make it back to earth. At the same time, Swigert knew, everything they had accomplished so far would be for naught if *Odyssey* didn't function properly when they got there. *Aquarius* was their lifeboat, but only *Odyssey* could ferry them through the earth's atmosphere to splashdown. Right now the command module was nothing more than an inert shell. Swigert called it the "upstairs bedroom." No one had ever tried to revive a dead command module after more than three and a half days of forced hibernation in space; mission control would have to invent the necessary procedures. Swigert had already spent some time

---

For the first time since the accident, Jack Swigert's spirits lifted. Every second brought them closer to home instead of farther away from it.

---

thinking it over in his own mind—first align the computer, then get the entry monitor system going. . . . But the checklist was far too massive for him to devise on his own. He would need to have it in writing, carefully scripted, down to the last circuit breaker. And it would have to work right the first time; there would be no postponing reentry. Swigert knew only too well that under normal circumstances it took three months to write a reentry checklist, test it in the simulator, and work the bugs out. The people in Houston had less than two days left to finish this one.

But in the back rooms of Building 30, down the hall from mission control, the reentry checklist was just one of the many efforts which consumed engineers and mission planners. Ken Mattingly was in the thick of it, going around on a freelance basis, wherever the action was and wherever he could be of use. In each room, teams of specialists huddled over schematics and scribbled calculations on blackboards, facing problems like how to conserve power, or water, or how to align the guidance system without a computer. For more than three years, they had been thinking of "what-if" situations and what to do about them. To Mattingly, the most amazing thing was that nearly every solution the teams were coming up with had already been thought of, and sometimes even tested, in previous missions. Firing the LM's engines while it was docked to the command module, for example, was one of the engineering tests crammed into the ten-day Apollo 9 flight.

Then there was the carbon dioxide problem. The LM was equipped with round canisters of lithium hydroxide to scrub exhaled carbon dioxide out of the cabin air. But there were only enough canisters to handle two men for about two days—not three men for more than three days. Carbon dioxide buildup would kill them as as surely as anything else. First would come bad headaches; then their hearts would begin to race. Finally they would grow drowsy and drift off into a sleep from which they would never wake up.

The command module had plenty of canisters, but they were square, the wrong shape for the LM's environmental control system. Unless some way could be found to use command module canisters in the LM, Lovell's crew would suffocate from their own breath. Someone remembered a simulation from Apollo 8 in which the cabin fans had failed, leaving the air inside the command module dangerously stagnant. And the engineers had figured out that in principle it would be possible to build a makeshift air purifier using only items that were readily available in the spacecraft—a plastic cover from a checklist book, a plastic storage bag, and so on. Now, eighteen months later, they would have to build this contraption for real, and make it work in an entirely different spacecraft. Then they would have to figure out how to describe it so the

The Apollo 13 astronauts' plight became front-page news, as typified by this headline from *The Topeka Daily Capital.* Before the emergency began, the progress of the flight had gone relatively unnoticed.

crew could build it without a blueprint or a picture of the finished product. That alone had kept a team from Deke Slayton's Crew Systems division busy since the first hours after the accident. It was one thing to have the pieces of Apollo 13's recovery in hand; it was another to put them together and make them work.

On the rare occasions when Mattingly stepped outside, he saw the whole space center lit up with activity. The lights were on for twenty-four hours a day in the offices of anyone who had anything to do with Apollo. At the Grumman plant on Long Island, North American in Downey, and contractors around the country, people showed up at their offices after the first reports of trouble, and they did not leave. Around the world, Apollo 13 had drawn the attention of more people than any other space mission except the first lunar landing. Headlines from Toronto to Tokyo blared the message that Apollo was in danger. News bulletins cut into radio and TV programs to report new developments. Crowds gathered anywhere there was a television set. In Rome, Pope Paul VI prayed for the astronauts' return before an audience of ten thousand people. At a religious festival in India, pilgrims numbering ten times as many offered prayers for the astronauts. Jim Lovell had no way of knowing that when he'd imagined Apollo 13 would be a mere footnote to history, he'd been very wrong.

10:16 P.M.

"We've gone a hell of a long time without any sleep," Lovell said to his crew, his voice thick with fatigue. Still on hot-mike, he added, "I didn't get hardly any sleep last night at all." After the PC + 2 burn there had been much to do, including getting the joined craft back into a thermal-control spin, still a difficult and frustrating task. Now that was done, but there were other things to worry about. He and Haise were powering down the LM, and at mission control's request, they were about to turn off the guidance system and the computer. It was the last thing he ever imagined he would do intentionally aboard a spacecraft heading for the moon. When the time came to pull the circuit breaker he double-checked with Capcom Vance Brand. If there were any more midcourse burns to make, Houston would have to come up with a way of making them without the computer. When Brand said they were working on just that, Lovell kept his skepticism to himself.

Swigert, meanwhile, was doing some worrying of his own. Which control system, he asked Brand, would Houston want him to use for reentry?

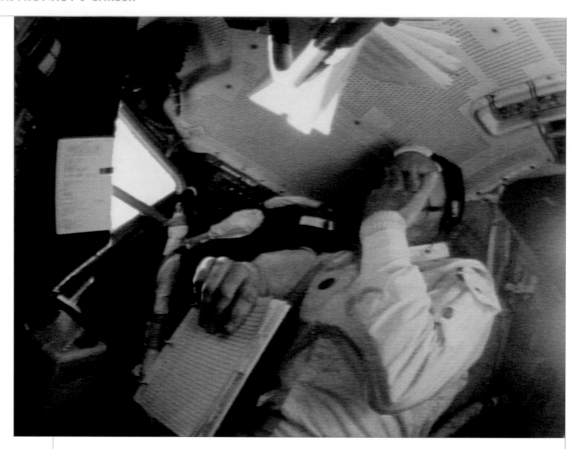

Mission control had asked him to make a check on one of *Odyssey*'s electrical buses—did they think it was okay? Suddenly Swigert heard a familiar voice.

"We think you guys are in great shape all around," Deke Slayton said. "Why don't you quit worrying and go to sleep?"

"Well, I think we might just do that," replied Jim Lovell, his tone suddenly more energetic. "Or part of us will."

WEDNESDAY, APRIL 15
12:44 A.M., HOUSTON TIME
3 DAYS, 11 HOURS,
31 MINUTES MISSION
ELAPSED TIME

The moon had dwindled steadily since the PC + 2 burn; now, some 19,000 miles away, it was about the size of a golf ball at arm's length. But that was still close enough to see craters with the unaided eye, and Fred Haise, alone in *Aquarius* while his crewmates slept in the "upstairs bedroom," took some time to look at it. Near the terminator, just emerging from night, he could barely make out the highland region called Fra Mauro, where he and Lovell were to land. And if it was a strange and alien place, then it had become only too familiar to Haise. As backup to Buzz Aldrin on Apollo 11 he'd mastered the technical and physical demands of getting to the surface of the moon and working there. By the time he began preparing for Apollo 13, Haise was a trained geo-

logic observer who was as excited about the science of his mission as the flying. But the moon beyond *Aquarius*'s windows was the face of disappointment.

At least he could feel good about *Aquarius.* Right now, with only the environmental control system and the radio going, the lander was drawing just a bit more than 12 amps—far less than Haise had figured on. When this flight was over, people would really know what a lunar module could do. *Aquarius* was the reason Haise had been able to sleep for five hours the morning after the accident. He'd trusted his calculations, and it turned out he'd been right. Lovell hadn't found it as easy to accept Haise's results, but Haise knew commanders tended to worry.

With almost nothing to do himself—at the moment, his only job was switching from one antenna to another as the spacecraft slowly turned—Haise's thoughts turned to Houston. From what the Capcoms had been saying, he could just imagine the massive, round-the-clock effort by the flight controllers and engineers. Apollo 13 had become their mission, their test, far more than it was his. According to Jack Lousma, things were going so well that they were already planning for splashdown. He asked Haise, "How would you like to spend a week on an aircraft carrier getting back?"

"If I can get on an aircraft carrier," Haise answered, "I don't care how long it takes, Jack."

III: THE CHILL OF SPACE

"Fred is being relieved now. He went back to get some rest. This is Lovell here who's got the duty."

"Gee whiz," said a surprised Jack Lousma, "you got up kind of early, didn't you?"

In fact, Lovell had returned to *Aquarius* after little more than 3 hours. In part, it was because he knew Haise was tired; in part, it was because he could not stop thinking about what lay ahead. But mostly, it was because it had become very cold in the command module. Yesterday morning, when he and Swigert had tried to sleep there, sunlight had shafted through the windows, keeping them awake, and they had put up the window shades. Normally that wouldn't have been a mistake. But with *Odyssey*'s systems shut down the shades removed the only other source of heat—sunlight—from the cabin. Now it was as wet and cold as a damp cellar. Moisture from the men's breath and perspiration had condensed into a thin film of water droplets all over the walls, the instrument panel, the windows. Lovell tried to rest there after midnight Wednesday, but the chill penetrated the flimsy fabric of his sleeping bag. It was remarkable, though, how weightlessness affected things: He found that if he stayed perfectly still, his own body heat built up around him like a tenuous blanket, because there was no convection to carry it away. But his slightest motion dispelled this warmth and let in the cold. Finally he gave up trying to sleep.

Back in *Aquarius,* Lovell had something new to worry about: Jack Lousma had informed him that Apollo 13 was straying off course for reentry. As a veteran of one lunar reentry, Lovell knew only too well the narrow cone of safety the command module would have to fly through. Something was making Apollo 13's flight path too shallow. Lousma said a midcourse correction burn was planned for about 9 P.M. Wednesday evening. Lovell couldn't imagine how he would align the spacecraft properly for the burn with no computer or guidance system, but Lousma said they were working on it.

Even the carbon dioxide problem hadn't escaped Fred Haise when he did his calculations on *Aquarius*'s consumables. But when he totaled up the number of lithium hydroxide canisters and decided there were enough, he only figured on two men; he forgot that Swigert had to breathe too. Fortunately the engineers on the ground had been on top of the problem, and they'd radioed word about their plan to have Haise and his crewmates construct an air purifier. It would be like building a model airplane, they said, but it would end up looking more like a mailbox. By Wednesday morning, Lovell and Swigert had gathered together the necessary materials: two lithium hydroxide canisters from *Odyssey*, a couple of pieces of cardboard from a flight plan book, a couple of plastic storage bags, and the stuff Haise called space-age baling wire, gray tape. While Haise slept in *Odyssey*, Lovell and Swigert listened as Capcom Joe Kerwin described the procedure.

"Now then," Kerwin began, "we want you to take the tape and cut out two pieces about three feet long, or a good arm's length, and what we want you to do with them is make two belts around the sides of the canister. . . . " It was an elegant creation. Lovell and Swigert were told to encase each canister in one of the plastic bags, affixing the bag with tape. Before that, however, they were to use the cardboard to fashion a little archway over the top of the canister to keep the bag from getting sucked against the canister's inlet screens. "The next step is to stop up the bypass hole in the bottom of the canister. . . . We recommend that you either use a wet-wipe or cut off a piece of a sock and stuff it in there. . . . " With all the multimillion-dollar equipment aboard Apollo 13, survival had come down to cardboard, gray tape, and socks.

It took about an hour for Lovell and Swigert to finish two of the makeshift air purifiers, and they did resemble a mailbox—in fact, they looked exactly like the air purifier that astronauts and engineers had put together on earth. Later, after Haise installed one of them on the end of an oxygen hose, the carbon dioxide level fell toward normal. Looking back on it, the ingenuity and teamwork embodied in that "mailbox" would amaze him, as much as anything on the flight.

1:34 P.M.

"Jack, for your information," said Joe Kerwin brightly, "FIDO tells me that we are in the earth's sphere of influence and we're starting to accelerate."

It was a piece of news Swigert was glad to hear. "I thought it was about time we crossed," he told Kerwin. "Thank you. We're on our way back home."

In Houston, Gene Kranz and his White Team of controllers were racing the clock to come up with the reentry checklist. Nothing would draw more

Deke Slayton displays a makeshift air purifier for, left to right, flight director Milt Windler, mission planner Bill Tindall, flight operations director Sig Sjoberg, Chris Kraft, and Bob Gilruth. Intended to prevent a deadly buildup of carbon dioxide in the lunar module, the device could be assembled aboard the spacecraft from materials on hand.

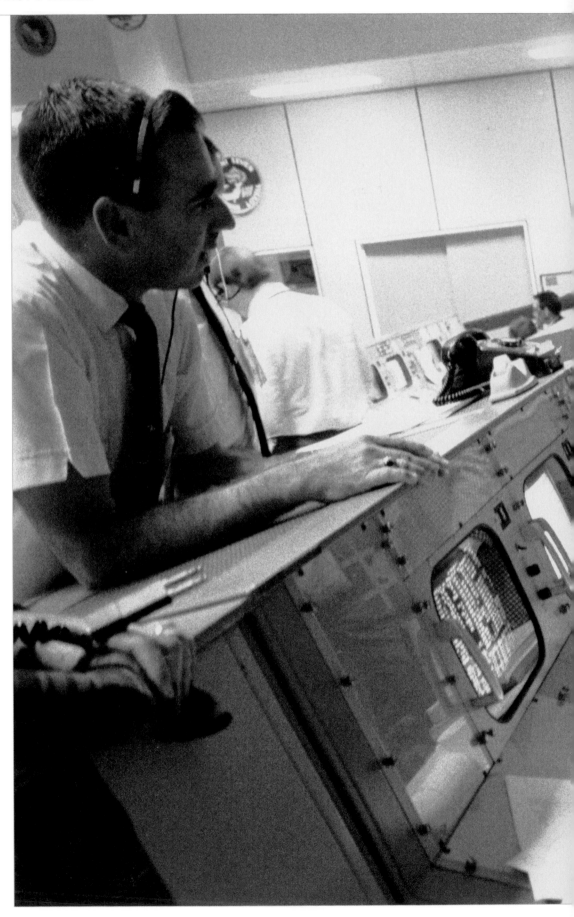

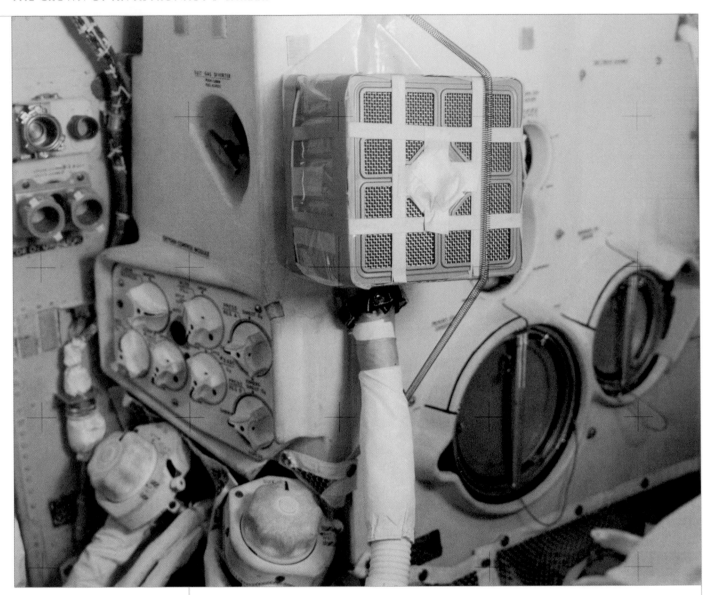

The completed carbon dioxide scrubber, attached to a ventilation hose inside *Aquarius,* looks exactly like the prototype constructed on earth. Minutes after the crew installed the device, carbon dioxide levels fell toward normal.

concentrated effort. Kranz split his people into three separate groups. One would write instructions for everything that should happen in the last four hours of the flight: powering up the command module, casting off the dead service module, letting go of the LM, and finally the reentry itself. Another group was to take care of translating the checklist into a form the astronauts could use. A third team would make sure that nothing called for in the checklist exceeded the slim reserves of battery power and water. At the top of their list was the daunting requirement to make the command module's batteries last three times longer than their normal 45 minutes. It was all but impossible—but they would have to figure out some way. Before long the White Team was aided by a "Tiger Team" of off-duty flight controllers. And Ken Mattingly hunkered down with them. They didn't have any time to waste; reentry was only 2 days away.

It was time to put Apollo 13 in position for the midcourse correction. Mission control had instructed Lovell to orient the spacecraft so that the sun shone directly through the LM's small overhead window. Then, looking through his forward-facing window, he would spin the stack around until he saw the earth, now a blue and white crescent, directly ahead. Then he would sight on the earth through a special gunsight, normally used for rendezvous maneuvers, and turn the spacecraft until the horns of the crescent were sitting on the crosshairs. Then, according to mission control, he would be in position to make the burn.

Earlier, when Lovell heard about this technique, he had a flash of recognition. Someone had dreamed it up during a "what-if" session before Apollo 8. Since then it had been taken out of the flight plan books as unlikely, and had been long forgotten—except, that is, by the experts in mission control. Lovell never would have dreamed, back then, that he would have to stake his life on such a scheme.

Around 9 P.M. Haise copied data for the burn; almost 90 minutes later Lovell had wrestled the stack into position. He informed Houston that he was ready for the burn and added that he hoped the guys in the back room knew what they were doing. When the moment came to fire *Aquarius*'s descent engine, all three men played a part. Lovell turned on the engine and controlled the LM's roll; Haise kept the stack oriented in pitch. And Jack Swigert kept his eye on the clock and called out the start and stop times. Fourteen seconds after it began, the midcourse burn was over.

"Nice work," radioed Jack Lousma.

"Let's hope it was," Lovell answered.

Lovell was alone on watch while Swigert and Haise slept. There would be no sleep for Haise in the cold atmosphere of *Odyssey*—they had started calling it "the refrigerator"—so he had brought down one of the sleeping bags. Now he was a strange sight, wrapped in the bag, his weightless body suspended upside down with his feet up in the tunnel and his head above the engine cover. To keep from floating away, he'd hooked the loop on the bag's zipper to

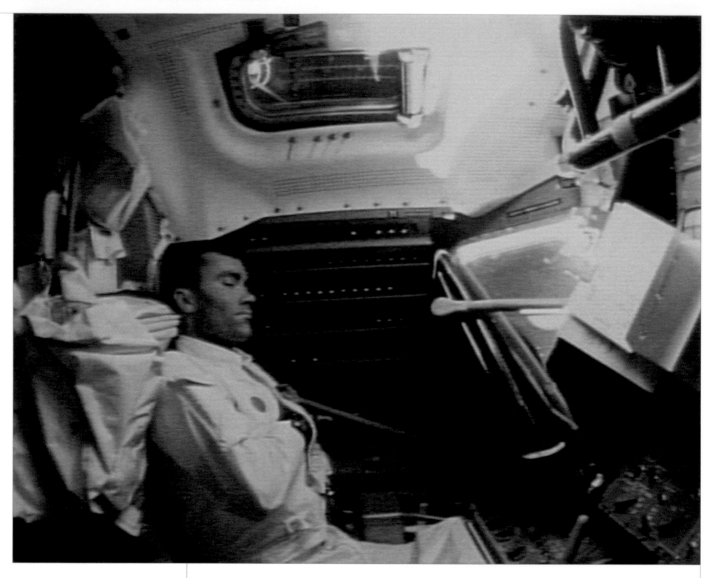

Fred Haise catches a few winks of fitful sleep in the lunar module's chilly cabin. During the voyage home Haise contracted a kidney infection brought on by unsanitary conditions in a makeshift urine-collection system.

the hatch handle. Swigert was down on the floor just beneath the instrument panel; he'd wrapped a restraint harness around his arm to anchor himself.

Lovell had found that he could close his eyes and doze off for 15 minutes, floating before *Aquarius*'s controls, and wake up feeling a little refreshed. He didn't feel nearly as exhausted now as he had at the end of those 20 hours in lunar orbit on Apollo 8. The difference was adrenaline; the constant demands of keeping his crew and his spacecraft functioning kept him going.

At least now, the end was in sight. From mission control, Jack Lousma had good news about *Aquarius*'s water: It was holding out so well at the present rate that there was enough to last 20 hours beyond reentry. "Your luck is really hanging in there," Lousma said.

Luck: Lovell had to smile at the irony of the word. He wasn't a superstitious man, and so when he found out he'd be flying Apollo 13 he didn't think anything of it. In fact, some of his Italian friends were happy for him; they

Jack Swigert emerges from the tunnel linking *Aquarius* with the command module *Odyssey*. The astronauts jokingly referred to their two-room quarters as the first space station.

said 13 was a very lucky number. So much for lucky numbers. And even though, after the accident, he'd wondered, *Why me?*—just as Swigert must have—he had to concede that, after 572 hours in space, he was a good target for the law of averages. What he hadn't realized at the time was how lucky he'd really been. After the accident he'd told Swigert and Haise, "It couldn't have happened at a worse time." After all, the explosion occurred just at the point where Apollo 13 was too far from earth to turn around. But he was wrong, Lovell knew now. If the explosion had happened in lunar orbit, getting out of orbit with only *Aquarius*'s engines would have been a much iffier proposition. If it had happened when the LM was on the surface, there would have been no recovery. He and Haise would be doomed men. And Jack Swigert would be dead, circling the moon in a command module without power, or an engine, or oxygen. Lovell realized it couldn't have happened at a better time.

Unlike Neil Armstrong and Pete Conrad, Jim Lovell had not received any pre-flight promise from Tom Paine about letting him and his crew fly the next mission if they had to abort. Another team of astronauts would have that chance. But when would that be? At the very least, Lovell suspected, there would be a hiatus of months while NASA tried to figure out what went wrong. At worst—if they didn't make it back—it could mean the end of the Apollo program. Yesterday, during an idle moment, Lovell had said to Jack Swigert, "I'm afraid this is going to be the last lunar mission for a long time." What he didn't know then was that the comment was heard live, on hot-mike. Tom Paine, waiting out the ordeal, went before the press to explain that, yes, there would be an investigation into Apollo 13, but after that, most definitely, NASA was going back to the moon.

● ◑ ◐ ○ ○ ○ ○ ◑ ◐

By Thursday afternoon, the cold had begun to invade *Aquarius*. The command module, again, was even colder; at one point Lovell went up to *Odyssey* to get some hot dogs out of the pantry, and they were practically frozen. The men had donned a second set of cotton long johns, and they discussed the idea of putting on their space suits to stay warm, but had decided against it. They wouldn't be able to turn on the suit fans, because they ate up too much power. Without any ventilation they'd overheat and perspire, then run the risk of getting seriously chilled if they had to take them off. But their long johns provided little warmth, and the Teflon-coated fabric of their coveralls was cold to the touch.

Jack Swigert, meanwhile, had his own troubles. Almost a day earlier he'd been in his usual place, straddling the ascent engine cover, when he realized his feet were immersed in a puddle of water; the LM's drinking-water dispenser had been leaking. Swigert's feet were soaked, and soon, after several trips back and forth to the command module, half-frozen. He'd been rubbing them ever since to get them dry. This morning, Lovell and Haise had broken out their lunar boots to keep their own feet warm, and Haise had offered his pair to Swigert. But he declined; he wanted to keep rubbing his feet.

The afternoon dragged on with the three men shoulder to shoulder in Aquarius, trying to stay warm.

The afternoon dragged on with the three men shoulder to shoulder in *Aquarius,* trying to stay warm. Mission control had promised to read up the lengthy checklist for tomorrow's reentry, but it wasn't ready yet. The cold only made the waiting seem longer.

Around 6 P.M., with no checklist yet in sight, the conversation in *Aquarius* turned to Ken Mattingly. If he were sick, he ought to be breaking out with red spots by now. Before the flight Lovell had worked out a code with the Capcoms so he could find out Mattingly's condition covertly. Now he asked Vance Brand, "Are the flowers in bloom in Houston?"

"No, not yet," Brand replied. "Still must be winter."

"Suspicions confirmed," Lovell said.

"I doubt if they will be blooming even Saturday when you return," Brand ventured.

Not only was Mattingly healthy, but he'd been working like mad with Gene Kranz's controllers to put together the reentry checklist. At 6:30, Vance

Brand radioed that it was nearly ready and that Swigert should get ready to copy it down. "He'll need a lot of paper," Brand said. By 7:19, Swigert was still waiting, and Lovell could no longer hide his irritation. They couldn't just wait around up here, he told Brand; they had to have the procedures in time to study them, "and then we've got to get the people to sleep." Finally, around 7:30 P.M., Mattingly entered mission control, checklist in hand, and sat down at the Capcom's console to talk to the man whose fate had been so strangely entwined with his own. No words of greeting passed between them, no ironic remarks. It was all business.

"Hello, *Aquarius;* Houston. How do you read?"

"Okay. Very good, Ken."

"Okay. Let me take it from the top, here." There followed a conversation that lasted the better part of 2 hours, as Mattingly read every switch setting, every keystroke of the computer, the steps that would bring *Odyssey* back to life and ready for reentry. After every line, Swigert repeated what he had heard. Around 9:15, it was time to copy the LM checklist. Mattingly assured Swigert that both checklists had been tested together in the simulators, and he added, "We think we've got all the little surprises ironed out for you."

"I hope so," Swigert said, "because tomorrow is examination time."

"Fred, are you sleeping?"

"Go ahead."

Mission control had planned to let the crew of Apollo 13 sleep through the night. But early Friday morning Haise was awakened by Jack Lousma, who had a minor change in the checklist for him to copy. Soon it was clear that Jack Swigert was also awake. For the benefit of the doctors, who were keeping tabs on the astronauts' rest periods, Lousma asked, "How much sleep did you get, Jack?"

"I guess maybe two or three hours."

"You plan to get any more?"

"Well, if I get everything done, I'll try, but I tell you, it's almost impossible to sleep. All of us have that same problem. It's just too cold. . . . "

By now, *Aquarius* too was like a damp cave. The temperature hovered in the mid-forties. The LM's environmental control system, designed for warmer temperatures and only two men, was overloaded with moisture, not only from the exhalations of the three men, but in the wet, frigid air that

Lovell, Swigert, and Haise float shoulder to shoulder in *Aquarius.* By this time, the men were suffering from lack of sleep, cold, and dehydration.

drifted in from *Odyssey.* Big globs of water, shimmering in zero g, clung to every bend in the exposed plumbing and the wire harnesses. The windows were clouded with moisture. The men had to wipe off the instrument panel to read a gauge. The stack had strayed so far from its normal orientation—the thermal-control spin had deteriorated badly—that the sun no longer appeared in *Aquarius*'s windows; it was shining down on the service module's engine bell. The men waited out these last hours in near-darkness, studying the checklist by flashlight, rubbing their hands together.

Years later, Ken Mattingly would look back on the performance of Lovell and his crew with amazement. For more than three days they had been living only for survival, in miserable conditions, with nothing exciting to break up the long hours of waiting. And yet, they never lost their temper with each other, or with mission control. And if Jim Lovell got on the radio every once in a while and let his frustration hang out a little, he was still the epitome of restraint. If Apollo 13 had to happen to any spacecraft commander, Mattingly would say later, there wasn't anyone could have handled it better than Jim Lovell.

2:06 A.M.

"Hey, Jim, while you're up and things are nice and quiet, let me give you a couple of things to think about. . . . " Deke Slayton sounded almost as tired as the man he was talking to. "I know none of you are sleeping worth a damn because it's so cold, and you might want to dig out the medical kit . . . and pull out a couple of Dexedrines apiece. . . . "

"We might consider it," Lovell answered; privately he worried that the letdown from the drug would leave them more tired than they already were.

"Wish we could figure out a way to get a hot cup of coffee up to you," Slayton said. "It'd probably taste pretty good about now, wouldn't it?"

"Yeah, it sure would. You don't realize how cold this thing becomes. . . . "

Hearing this, Jack Lousma offered his own words of encouragement: "Hang in there. It won't be long now."

2:35 A.M.

"Okay, Skipper. We figured out a way for you to keep warm." It was Jack Lousma's voice. "We decided to start powering you up now."

"Sounds good," Lovell said, "and you're sure we have plenty of electrical power to do this?"

"That's affirmative." In fact, *Aquarius* had come through so well that from here on in, there would be twice as much power and water as the men would need. Lousma's go-ahead came none too soon; as he set to work turning on the systems, Haise looked at his window and thought he could almost see frost.

Haise filmed the frigid cabin of the command module, whose systems Swigert had shut down soon after the accident. In the lower-left corner is the top hatch, removed for access to the lander; the side-hatch window, clouded by condensation, appears in the upper-right corner.

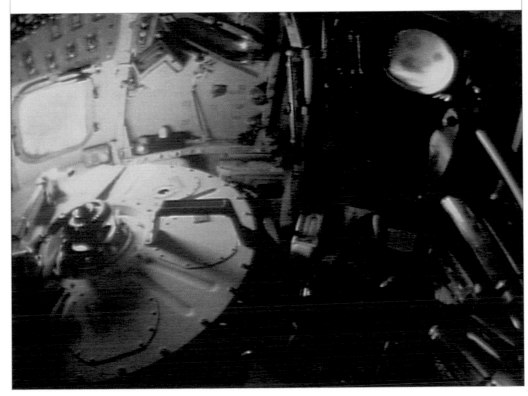

Earth—the "grand oasis" in Jim Lovell's words—beckons the astronauts as they near the end of their ordeal. Baja California is visible through the hole in the cloud cover near the center of the photograph.

By 3:15, Lovell and Haise had brought *Aquarius* fully to life, and Lovell told Lousma it was warming up a little. Lousma answered, "Duck blinds are always warmer, Jim, when the birds are flying."

Apollo 13 was only 55,000 miles away from earth, heading toward it at 6,100 miles per hour. When Lovell steered the stack into position for a new mid-course burn, the planet was a crescent of impressive size and growing larger by the minute. But even now, Lovell's crew could not rest easy. Something was still causing their angle of attack to shallow slightly. Throughout the past couple of days, the men had seen more spurts of gas from the service module—suddenly the constellations would be lost in a flurry of sparkles—but Houston assured them that couldn't be the cause. Whatever the force bending Apollo 13's trajectory, it had to be corrected. All that was needed was a gentle, 21-second nudge from *Aquarius*'s maneuvering thrusters. But Lovell was so tired that he called up the computer program for the descent engine instead. In Houston, sharp-eyed controllers caught the mistake in plenty of time.

Tiredness was not the only cause of Lovell's mistake. Ever since Houston had told him that water was tight—confirming Haise's numbers—Lovell had decided to ration drinking water. The men had made good use of the water left in the command module's drinking tank, but Lovell had been particularly stringent about his own water consumption. The situation was compounded by the fact that in zero g, astronauts generally don't feel thirsty as often as they do on the ground. Though he did not realize it, he was becoming dehydrated. He would later find out that the loss of electrolytes made him more prone to errors.

Fred Haise had his own problem: he was sick, and all because of a misunderstanding with mission control. After the accident, with *Odyssey* powered down, the men couldn't use the normal system of dumping urine overboard; it required electricity to heat the line and keep it from freezing. Instead, they'd used an alternate line, which dumped urine through a fitting in the command module's side hatch. But mission control told them that the resulting swarm of ice droplets around the spacecraft was ruining their Doppler tracking data; please don't dump any more urine. They neglected to say, "for the time being." Lovell and his crew interpreted the request as a permanent ban. The misunderstanding created an immediate problem of where to store urine, and for nearly three days they'd been using every suitable bag and container, stashing them in the LM's storage cabinets. But it also led, indirectly, to Haise's illness. Mission control's request came at a time when the men were very busy. The easiest way to store urine, they decided, was to use the collection bags they normally wore inside their space suits. They kept them on while they worked, for some number of hours, until the pace of events slackened. That saved some time, but it also meant they were bathing themselves in urine—and that made them vulnerable to infection. By Thursday evening, Haise had developed a urinary tract infection, although he did not know it. He knew only that he was experiencing a burning pain when he urinated. And on Friday morning, when Haise went up to *Odyssey* to go to the bathroom, he bumped up against the walls with his bare skin. The cold metal drained the warmth out of him. When he returned to *Aquarius* he was shivering badly. He zipped himself into a sleeping bag and floated before the instrument panel. Swigert felt his head; he didn't seem to have a fever. The shivering stopped after a couple of hours, but for the first time in the flight, Haise felt truly exhausted. ◖

7:15 A.M.

After the midcourse burn, it was time to cast off the stricken service module. Jack Swigert had been in *Odyssey* for more than two hours, where he had begun to turn on some of the systems, drawing power from *Aquarius*—another bit of ingenuity from Houston. To guard against a lethal mistake, Swigert had put tape over the switch marked "LM JETTISON." Down in *Aquarius,* Lovell pulsed the thrusters to impart momentum, and yelled, "Fire!" Swigert hit the switch marked "SM SEP" and there was a bang of pyrotechnic bolts, and the service module was away. Lovell immediately swung the joined LM and command module around—with the massive service module gone, the craft suddenly responded crisply to his commands—and through the overhead window he saw the service module, tumbling slowly as it departed. For the first time he could see what had happened to Apollo 13. ◖

"There's one whole side of that spacecraft missing! Look out there, would you?" Instead of the small puncture wound Lovell had imagined, the explosion had blown off an entire panel from the skin of the service module. Inside he could see tanks, pipes, and other equipment amid a tangle of torn Mylar insulation. At the base of the cylindrical module, they could see that the big communications antenna had been bent out of position. The men fired off pictures until the service module was a speck in the distance.

IN 1970

Midnight Cowboy wins best picture Oscar; John Wayne is named best actor for *True Grit.*

Biafran forces surrender to the Nigerian Army, ending civil war in Nigeria.

Erich Segal publishes *Love Story.*

Kareem Abdul-Jabbar is named Rookie of the Year.

In Houston, flight controllers were chilled by the verbal picture of destruction. As to the cause of the accident, that would be for the review board to determine. But Ken Mattingly had a good idea of what they would find; he had been directly involved. Back in March, the launch teams at the Cape had conducted the Countdown Demonstration Test, which included filling the spacecraft's tanks with cryogens and then emptying them again. When it came time to drain the service module's oxygen tanks, technicians were unable to remove the super-cold fluid from tank number 2. For eleven days, they studied the problem. Mattingly went to briefings by the engineers, who had worked around the clock to find out what was wrong. They explained that the tank had originally been scheduled for Apollo 10, but had been removed for modifications and replaced. Sometime after that, the tank had been accidentally dropped a distance of two inches—a minor jolt, but enough to damage the tube assembly used to fill and empty the tank. That problem, never completely fixed, was the reason for the present difficulties. But there was an alternate procedure, they said, that would work around the problem. They would turn on the tank's built-in heaters and warm the liquid oxygen until it boiled off through the tank's relief valve. They asked Mattingly whether that sounded okay; he agreed.

But neither Mattingly nor anyone else knew that the heater inside the tank had a design flaw. Back in 1965, the entire Apollo spacecraft, which was intended to operate at 28 volts of electricity, was upgraded to accept 65 volts from ground test equipment. Everything, that is, except the thermostat inside the service module's oxygen tanks. It was still rated for only 28 volts; no one caught the error. It was that thermostat that was supposed to shut off the heaters when the temperature inside the tank reached 80 degrees Fahrenheit. During the draining operation, the thermostat was activated, but the excess voltage caused an arc that welded its electrical contacts shut. The heaters stayed on for 8 solid hours, long enough for the temperature to reach 1,000 degrees. The technician monitoring the test was unaware of the dangerous condition because his temperature gauge went no higher than 85 degrees. Inside tank number 2, the intense heat baked and cracked Teflon insulation covering wires on a nearby motor; the motor controlled the fan inside the tank.

Nothing more happened until 2 days, 7 hours, and 54 minutes into the flight of Apollo 13, when Jack Swigert turned on the fan at mission control's request. At that point, the review board would later decide, there was an electrical arc inside tank number 2. In moments a fire had started inside the tank, fed by the generous supply of oxygen. Seconds later, the sudden surge of

pressure blew off the tank's dome, flooding the surroundings with oxygen. When the pressure reached 30 pounds per square inch, the panel blew away in a jolt that probably also severed the plumbing for tank number 1, crippling Apollo 13.

In the test flight business, big mistakes are often the result of many little ones. Mattingly would always carry with him the knowledge that he had signed off on Apollo 13's damaged oxygen tank. But he was by no means the only one who had failed to stop the process that led to the accident; a number of managers and engineers had also said, "Go ahead." So had Jim Lovell.

A strange and unprecedented combination, lunar module and command module, sped earthward. Inside *Odyssey*'s dark, frigid cabin Jack Swigert and Fred Haise faced the moment of truth. With just 2½ hours to go before reentry, it was finally time to bring the command module fully back to life. They knew the systems must be even colder than it felt inside the cabin. No one knew if the navigation system would function again after such a prolonged cold soak. The design specifications said it wouldn't work. They hoped the specs were wrong.

Before they could do anything they had to wipe off the gauges, because there was water all over the instrument panel. They knew there must be water *behind* the panel too, clinging to the insulation around electrical connections, perhaps ready to short-circuit as soon as they began pushing in circuit breakers. Swigert and Haise had no choice but to charge ahead, hoping they didn't see sparks or smell smoke. With relief, they realized their fears were unfounded. Undoubtedly the modifications made to both the command module and the LM after the Fire had prevented this potential disaster. And every one of *Odyssey*'s systems revived in perfect condition, even the computer. Swigert was so glad to have his command module back that he didn't even need Dexedrine.

But for Haise's friend, lunar module number 7, time had nearly run out. Swigert and Haise were in their couches in *Odyssey* while Lovell, now crowded in *Aquarius* among bags of unneeded gear, steered the joined craft into position. When Lovell was back in the command module, the men sealed the hatches between the two craft. When Swigert hit the switch there was a loud bang and a jolt far more violent than the oxygen tank explosion had been. *Odyssey* pitched perilously close to gimbal lock, but Swigert quickly steadied the ship. Then Haise watched the departing LM, legs extended, poised for landing on alien soil, but instead headed for a reentry it would not survive. He wished there had been some way to bring it home and put it in a museum. Joe Kerwin spoke for all of them: "Farewell, *Aquarius,* and we thank you."

11:43 A.M.

Odyssey sped through darkness toward its rendezvous with earth. Inside, Lovell, Swigert, and Haise were strapped in their couches, exchanging last-minute bits of information with Houston. After three days of constant background noise from *Aquarius*'s fans and pumps, the command module seemed unnaturally quiet. But now, with everything working perfectly, there was still a troubling unknown in Fred Haise's mind: the condition of the heat shield. Even before they saw the service module, they had talked among themselves about the possibility that it had been cracked by the explosion. Of course, that resin-filled honeycomb structure was extremely tough. And a small crack would probably seal itself as the heat shield began to ablate in the heat of reentry. A serious crack was another matter. But soon Haise was far too busy to think about it. ❨

Now, with just 7 minutes to go until reentry, Joe Kerwin radioed the welcome news that *Odyssey*'s guidance system was perfect. Swigert suspected that when it was all over the trajectory experts in mission control would have a great party, and he told Houston he wished he could go to it. Kerwin relayed an answer from someone in the control room: "We'll cover for you guys, and if Jack's got any phone numbers he wants us to call, pass them down."

With three minutes to go, Lovell and his crew saw a tiny and distant moon, and as they watched, it blinked out, sinking behind the dark horizon exactly when mission control had predicted. And then, the blackness of space gave way to the first, soft light of ionized gas, a glow known only to the returning space traveler.

TIMBER COVE, TEXAS

A crowd had gathered with Marilyn Lovell to watch the recovery on television. In the past three and a half days she had hardly left her house. Yesterday, she had lunched with Fred Haise's wife, Mary—who was very pregnant—at

Deke Slayton's house. And earlier this morning, she had gone to get Jim's mother from the nursing home so that she could watch too. For the past two days, her spirits rose steadily as the reports on Apollo 13 improved. Still, there had been difficult moments. She knew Jim wasn't getting much rest. At one point, someone sent word that Dr. Berry thought it might be a good idea if she went down to mission control and talked to Jim directly, to convince him to go to sleep. She passed word back that she just couldn't do that. She knew that if she tried to talk to him, she would fall apart.

Then, earlier this morning, she had been sitting by the squawk box when she heard Jim say, "There's one whole side of that spacecraft missing!" And she panicked. With no NASA people around to ask, she called Pete Conrad, who said, "I'll check on it, I'll check on it"—and soon she was reassured. But she would not rest easy until she saw the command module coming down on its parachutes.

Now Marilyn sat in front of the TV among her close friends, with little Jeffrey at her side. Just minutes ago she had heard the last conversations between Jack Swigert and mission control. "Everybody says you're looking great," Kerwin had said. "Welcome home."

"Thank you," Swigert had said. And then there was silence from Apollo 13. Long minutes passed as she waited anxiously for the communications blackout to end. She heard the NASA public affairs commentator say, "About thirty seconds to go for blackout. Less than ten seconds. We will attempt to contact Apollo 13 through one of the [Apollo Range Instrumentation Aircraft]."

And there was silence. Half a minute went by; nothing. Now it was a full minute. Suddenly: "We've had a report that the ARIA 4 aircraft has acquisition of signal." Now she heard Joe Kerwin's voice. "*Odyssey,* Houston. Standing by."

Long seconds passed. Finally, there was Jack Swigert: "Okay, Joe." And still, she could not relax; there were the parachutes yet to come. She was not alone in her tension; no one, either in mission control or inside the command module, knew whether the electronics to deploy the chutes were still good.

On television, the broadcast from the recovery ship showed morning in the South Pacific, a calm sea under a patchwork of fair-weather clouds. She heard Kerwin radio to Apollo 13 that the weather was good. The public affairs officer counted down for the deployment of the drogue parachutes. Less than 2 minutes; now 1 minute; now 30 seconds. For interminable seconds it was hard to tell what was happening. Then, the TV cameras found the command module: a small dark cone floating out of the clouds on three beautiful

An extraordinary gathering of moonwalkers joins the Lovell family to watch the coverage of Apollo 13's return. Seated, from left: Pete Conrad, Buzz Aldrin, Lovell's mother Blanch, Barbara Lovell, Jeffrey Lovell, and Marilyn Lovell. Sitting at rear is Neil Armstrong.

orange parachutes. The room erupted in cheers and Marilyn hugged Jeffrey so tightly that he cried out.

Meanwhile, at the space center, cheers and applause thundered through mission control. Everyone shook hands and lit up cigars—including Ken Mattingly, who smiled and shook hands with Chuck Berry. And inside *Odyssey*, Jim Lovell remembered his ton-of-bricks splashdown from Apollo 8 and said to his crew, "Gentlemen, be prepared for a hard landing." But *Odyssey* must have caught the crest of a descending wave, for it fell gently onto a calm Pacific Ocean.

As they waited for the swimmers to arrive, Lovell, Swigert, and Haise lay in their couches, amazed that even now, after the heat of reentry, *Odyssey* was still so cold that they could see their breath. When the swimmers opened the hatch, a great cloud of frosty air issued into the tropical morning. The chill of

Mary Haise *(near right)* rejoices as *Odyssey* heads for a safe splashdown, while Marilyn Lovell *(far right)* hugs four-year-old Jeffrey. Exclaimed Mary Haise, "Oh, how glorious."

space would linger inside *Odyssey* long after Lovell, Swigcrt, and Haise were safely aboard the aircraft carrier *Iwo Jima*, recounting their ordeal for the doctors, and savoring the fresh, warm air of earth.

● ◑ ○ ○ ○ ○ ◑ ●

Marilyn Lovell had never seen a party like the one that enveloped her house at splashdown; nor had she seen so many champagne bottles emptied. People were leaning against the wall, unable to move. Marilyn lost her voice, but not before she went out to greet the reporters who had kept a patient vigil through the past four days. She jokingly gave them the old standby—"Happy, thrilled, and proud"—and one of them asked her what the past four days had been like. "It's been a nightmare," she told them. "I have never experienced anything like this in my life and I never hope to experience it again." That wasn't the kind of talk usually acceptable for an astronaut wife, but now for the first time since Jim became a test pilot, she had no reason to hide anything.

● ◑ ○ ○ ○ ○ ◑ ●

The day after Lovell's crew splashed down Richard Nixon came to the space center and presented the Presidential Medal of Freedom to Gene Kranz, Glynn Lunney, and the other flight directors of Apollo 13. Tom Paine was there too, sitting on the stage, watching as Nixon praised the ingenuity and courage of the NASA and industry personnel who had saved the lives of the three astronauts. For many in the audience, Apollo 13 would always be NASA's finest hour.

But the irony was that even now, Paine's vision for the space program was crumbling. In March the White House had issued a statement saying that space would no longer hold such a high position in the list of national priorities. There would be no expensive new programs, especially not in the area of manned spaceflight. There would be no money spent on any effort linked to the goal of sending humans to Mars. By summer, NASA's goal of a space station and reusable space shuttle was coming under fire in the Senate, where Walter Mondale protested, "I believe it would be unconscionable to embark on a project of such staggering cost when many of our citizens are malnourished, when our rivers and lakes are polluted, and when our cities and rural areas are dying." He asked, "What are our values? What do we think is more important?" Despite this and other outcries, Congress approved the NASA budget by a slim margin at the end of July. The next day Tom Paine surprised his colleagues at NASA by resigning as administrator to take a management job in industry. He would leave to his successor the coming battles for support—not from Congress, but from Richard Nixon.

But it was already clear that Nixon had rejected the vision proposed by the Space Task Group. By September, only the space shuttle had any chance of being approved—and that was still uncertain at best. As presidential adviser John Ehrlichman told historian John Logsdon years later, inflation and other priorities were more important; the budget just wasn't big enough to do everything NASA wanted. Furthermore, Nixon was unwilling to pay for a Mars effort that would not bear political fruit until after he left office. But Apollo was another matter. After Apollo 13 some of Nixon's advisers, wary that a space disaster would do him political harm, had urged him to cancel the rest of the moon missions, but he resisted. According to Ehrlichman, the moon program held magic for Nixon for one reason: he liked heroes. To him the astronauts represented the best the country could produce. It was good for the nation, Nixon believed, to have heroes. ❆

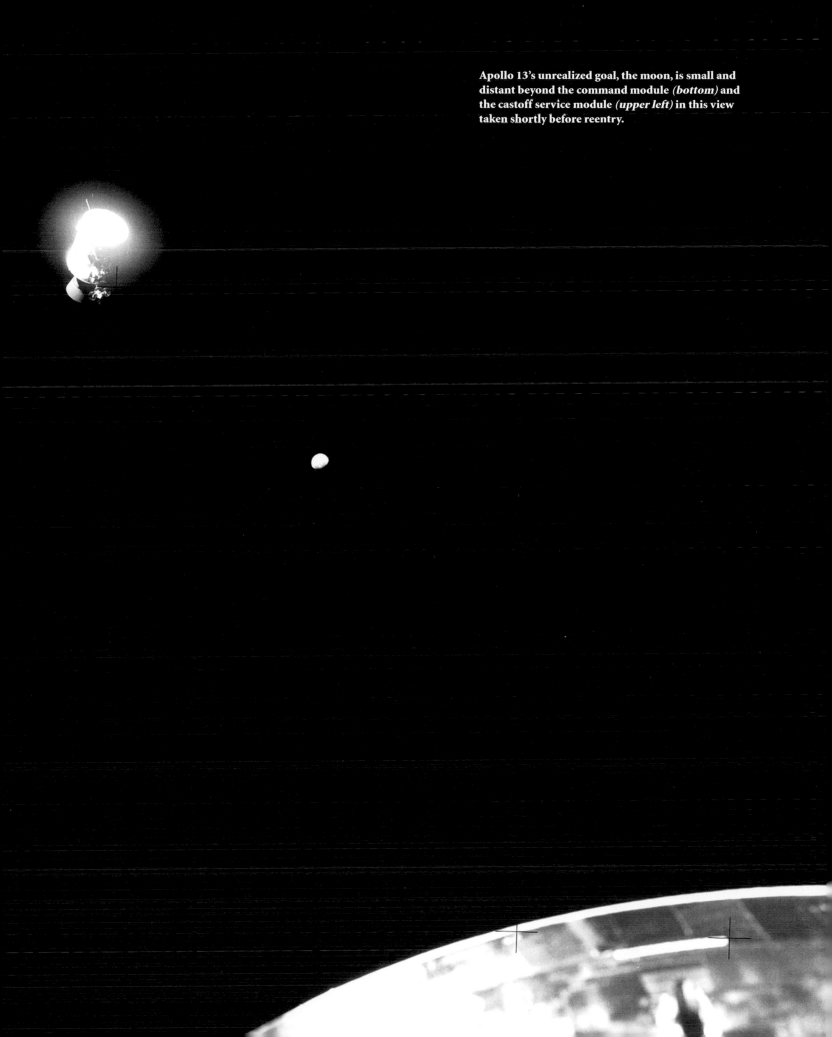

Apollo 13's unrealized goal, the moon, is small and distant beyond the command module *(bottom)* and the castoff service module *(upper left)* in this view taken shortly before reentry.

In a scene witnessed live by millions of anxious television viewers around the world, the command module *Odyssey* descends under its three main parachutes. Splashdown took place in the South Pacific, five days, 22 hours, and 54 minutes after liftoff from Cape Kennedy.

Haise, Lovell, and Swigert greet sailors and welcoming officials as they arrive on the deck of the U.S.S. *Iwo Jima*. Although drained by their ordeal, the astronauts suffered no long-term effects.

Against the backdrop of a hand-painted, welcome-home banner, Fred Haise *(center)* greets a throng of well-wishers gathered at his house to celebrate his safe return to earth from the harrowing voyage of Apollo 13.

CHAPTER 8

THE STORY OF A FULL-UP MISSION

APOLLO 14

I: BIG AL FLIES AGAIN

Some memories are so bright that the passage of time cannot dim them. For Alan Bartlett Shepard, the brightest was May 5, 1961: On that Friday morning, while a nation watched and listened, he lay inside a tiny spacecraft called *Freedom 7* atop a Redstone booster and waited to be hurled into space. The space program had suffered more than its share of failures, and images of exploding rockets were burned into the national psyche. The suspense was almost unbearable. This launch would give a badly needed boost to national morale and prestige—if Shepard wasn't killed. As for Shepard himself, he had a different worry: he didn't want to screw up.

Inside *Freedom 7,* Shepard fought nervousness by concentrating on his work. For hours the count was held up by one problem after another— first the weather, then technical difficulties—and Shepard's mood turned impatient. With less than three minutes to go, minor trouble with the Redstone halted the count yet again. Over the radio Shepard could hear the engineers in the blockhouse, anxiously debating whether to postpone the launch, and what he said was the stuff of folklore: "I'm cooler than you are. Why don't you fix your little problem and light this candle?" ☾

Alan Shepard trains for his Apollo 14 moonwalks in July 1970. For this, his second chance to fly in space, Shepard waited 10 years after his historic Mercury mission in 1961.

May 5, 1961: Shepard strides toward the Redstone rocket that will launch him on a 15-minute suborbital flight, making him the first American in space. He carries a portable air conditioner to provide ventilation within his space suit.

At 9:30 A.M. fire erupted from the Redstone and Shepard rose from the east coast of Florida and soared into a sunlit infinity. Fifteen minutes later, having arced like a guided cannonball to an altitude of 115 miles, *Freedom 7* splashed down in the Atlantic, and a nation rejoiced. Within days Shepard stood in the White House rose garden to receive NASA's distinguished service medal from John F. Kennedy. Afterward, he rode in an open car down Pennsylvania Avenue, to the adulation of thousands. Sitting next to him, Vice President Lyndon Johnson was amazed. "Where did all these people come from? You're a famous man, Shepard."

Yes, he was a national hero, but the most precious thing Shepard gained that day wasn't glory—the true pilot, he would say years later, never does it for the fame—but the satisfaction of having been first. Fierce competition for the first Mercury flight— and with it, the chance to become the first man in space—had been an undercurrent in everything the Original 7 did. But the choice was up to their boss, Bob Gilruth, the soft-spoken head of the Space Task Group who thought of the astronauts as "his boys." One day Gilruth asked each of the Seven to conduct a peer vote: "If you yourself cannot be the one to make the first flight," he said, "who do you think it should be?" Shepard wondered if this was some kind of stunt dreamed up by the shrinks who had been part of the Seven's lives since the astronaut selection. There was actually a duty psychologist, Bob Voas, assigned to them; Shepard never did trust him.

Shepard had no idea whether the peer vote was important, but he didn't worry about it. There were no bad pilots among the Seven, Shepard knew, although NASA had overlooked some discrepancies in their piloting backgrounds in its zeal to select perfect mental and physical specimens. But he had no doubts about where he stood—and never had. That supreme self-confidence was with him when he graduated from the Naval Academy in 1944, ready to end the conflict in the Pacific single-handed. After the war he began his aviation career, flying off carriers, and moved on to become a test pilot at Pax River. It was that quiet arrogance that Shep-ard, more than any of his colleagues, brought to the business of being an astronaut.

Still, Shepard doubted he would be chosen. He figured the first flight would go to his main rival: John Glenn. Among the Seven it was Glenn—charismatic and devout, with the Ohio small-town background and the sunny, freckle-faced look of an all-American—who most embodied the hero the country was looking for in its astronauts. At the Original 7's first press con-

---

Fifteen minutes later, having arced like a guided cannonball to an altitude of 115 miles, Freedom 7 splashed down in the Atlantic, and a nation rejoiced.

---

ference—which had scared some of them half to death—Glenn had been a natural. For an image-conscious NASA, Shepard figured, Glenn was the obvious choice. And Glenn, with his own résumé of test-flight accomplishments, was as determined as any of them to be first.

Around mid-January 1961, Gilruth came to see the Seven in their office and announced that after much consideration, the pilot for the first flight would be Alan Shepard. Grissom would get the second flight; Glenn would back up both flights. As Gilruth spoke, Shepard was looking at the floor, and long seconds passed before he looked up to see the other six coming over to congratulate him. It wasn't hard to guess that their smiles concealed great disappointment, but Shepard couldn't have known that John Glenn was so angry that he later considered going to Jim Webb to try to get the decision changed, then thought better of it. NASA kept the selection a secret, saying only that one of the three—Glenn, Grissom, or Shepard—would make the flight. ❮

Then came the tough part. Looking back, the months leading up to the flight had been a true test, especially putting up with the forecasters of doom. They warned that a man would not survive a space mission, that he would succumb to the g forces of launch and reentry, that weightlessness would wreak havoc on his vision, or his sense of orientation, or his bodily functions. Some psychologists were predicting that the astronaut would experience a profound sense of separation from earth—a "breakoff phenomenon"—and be rendered helpless. Nor were these pronouncements coming from the fringe. In April, representatives of the President's Scientific Advisory Committee went down to the NASA offices in Langley, Virginia, and voiced their trepidation about the flight. Shepard, Glenn, and Grissom all volunteered to take extra runs on the monstrous centrifuge at Johnsville, just to prove to the skeptics that their fears were unfounded. Meanwhile, within NASA, Mercury planners looked ahead confidently. A chimpanzee named Ham had already made the trip and had come back in good shape. However, because of some minor problems with the booster, NASA postponed Shepard's March launch until early May, to allow time for one more unmanned test. ❮

Then, early in the morning of April 12, Shepard's hopes of being the first man in space evaporated as the world learned that a twenty-seven-year-old Russian pilot named Yuri Gagarin had orbited the earth. Shepard fumed that NASA hadn't seized the chance to send him up in March, but there was nothing to do about it now. Shepard and everyone else connected with Project Mercury were caught up in preparations for the launch. After the May 2 attempt was scrubbed because of the weather, NASA could no longer conceal the pilot's identity. By the time he climbed atop the Redstone on May 5, the eyes of the nation were on him.

So were John Kennedy's. When Shepard flew, the president was already weighing the decision of whether to attempt a moon program. Shepard's flight put him over the edge. Shepard would always believe that the immense outpouring of pride that greeted his flight made a great impression on Kennedy. If a suborbital hop from Florida to Bermuda could so energize the nation, imagine what would happen if America put a man on the moon. By May 25, less than three weeks after Shepard's triumphant ride down Pennsylvania Avenue, Kennedy had made his decision. That day, as he addressed a joint session of Congress, his closest aides could tell he was nervous—by the way he kept playing with the pages of his speech, creasing them, smoothing them. They sensed that Kennedy wasn't sure he was doing the right thing, to ask for such

an enormous sum of money for something so audacious. But none of that showed in his voice as he spoke the words that would mark the genesis of Apollo.

When Shepard next saw Kennedy, it was in February 1962, after John Glenn's orbital mission, which drew so much public celebration that it overshadowed the reaction to his own flight. Aboard Air Force One, flying from West Palm Beach to Washington, Glenn, Shepard, and Grissom sat with Kennedy undisturbed for most of the 90-minute flight. They talked about Glenn's experiences, and about the moon program. Aides came in and out of the forward cabin, but Kennedy wasn't interested in them. He didn't want to talk about anything but space. That day, Shepard saw the man's pioneer spirit. Some said the moon decision had been motivated by political concerns, but to Shepard it had been an act of statesmanship, not politics. And it was

---

But that summer Shepard's fortunes changed abruptly. At home one day, he was suddenly swept with dizziness, nausea, and vomiting.

---

clear to him, inside Air Force One, that the lure of space exploration was very real to John Kennedy. The following September, when Kennedy delivered his stirring speech at Rice University stadium, Shepard saw a president who was behind Apollo all the way.

By that time, Shepard was already thinking about getting into space again. His own flight had been so brief—15 minutes and 28 seconds—and so busy—two minutes for this test, a minute for that one—that it had merely whetted his appetite. For raw thrills, nothing could top it, but it hadn't been long enough for him to really show what he could do. He wanted the next flight he could get, though he would have to wait his turn on the rotation. That came after Gordon Cooper circled the earth for 34 hours in May 1963; Shepard made a bid for a three-day mission. But Jim Webb, anxious to get on with Gemini, was very much against prolonging Mercury, with the time, energy, and expense that would entail. With Webb's knowledge, Shepard took his case to Kennedy himself, but in the end, Webb prevailed.

Shepard's disappointment was short-lived. He was still at the top of the rotation. Gemini was coming, and after that, he could look forward to Apollo. Soon Shepard was assigned as the commander of the first Gemini mission, with Tom Stafford as his copilot. But that summer Shepard's fortunes changed abruptly. At home one day, he was suddenly swept with dizziness, nausea, and vomiting. When he recovered, he thought he was coming down with the flu, but within days he had a second attack, and then a third. Reluctantly he took his problem to the NASA doctors. Their diagnosis was Ménière's syndrome, an inner ear disorder characterized by a buildup of excess fluid in the semicircular canals, causing impaired balance and attacks of vertigo. Medications would control the vertigo, the doctors said, but the prognosis was bad. There was no cure for Ménière's syndrome, and unless Shepard was among the 25 percent for whom the disease clears up on its own, he would never make another spaceflight. At first, Shepard was optimistic; he told Stafford, "We're going to fight this thing." But soon the pair were taken off the flight. Stafford was eventually reassigned to a Gemini mission, but Shepard had no recourse. He was forbidden even to fly an airplane by himself. For an astronaut, it was like a death sentence.

For six years, Shepard lived with the stigma of being grounded. He never gave any thought to leaving the astronaut corps. He still hoped the disease might go away spontaneously. In the meantime, Deke Slayton had his hands full trying to run the newly created Flight Crew Operations Directorate, and he asked Shepard to take over the Astronaut Office. Through Gemini and into Apollo, Shepard handled the humdrum administrative affairs, coordinating training and travel schedules, approving interview requests, and the like. Knowing the collection of egos under him, the prima donnas who balked at serving on a backup crew, Shepard felt compelled to use a firm hand. It meant he had few friends in the office, but that never seemed to bother him; in fact, it was better for morale not to give even the appearance of favoritism.

A Chief Astronaut who could not fly was like a warrior-king who could not fight, and Shepard's frustration was magnified by his role in preparing other astronauts for their missions. It began with the secret crew-selection process, of which Shepard was always a key part. Then, as each crew trained, he followed their progress at the Cape. On launch morning he ate breakfast with them, and then he did the hardest thing of all: he walked them out the door of the crew quarters and watched as they climbed into the transfer van and rode away, heading for the launch pad.

But the care and feeding of astronauts did little to satisfy Shepard's appetite for a challenge. To fill the void, he turned to business. Even before he was an astronaut, Shepard had dabbled in real estate and made a respectable sum of money. The Mercury era had left him well connected to the Houston

social and business scene, and he was getting some expert advice from Leo De'Orsey, the lawyer who guided the Original 7 through the *Life* magazine contract and other unfamiliar waters. Shepard logged his share of failures, but also successes, especially with his lucrative part-ownership of Houston's Bay Town bank. Within a few years he had amassed a small empire that at one time or another included partnerships in shopping centers, hotels, and other assorted enterprises around the country. He bought a $200,000 house in Houston's opulent River Oaks district. If Shepard's escapades drew some private envy and resentment both inside and outside the Astronaut Office, no one could fault him for the way he looked after the astronauts' interests within the space center. And while he was sometimes known to put in half-days in the Astronaut Office and spend the rest on outside business, one of his Mercury colleagues would later say, "Al never let anybody down in that job.

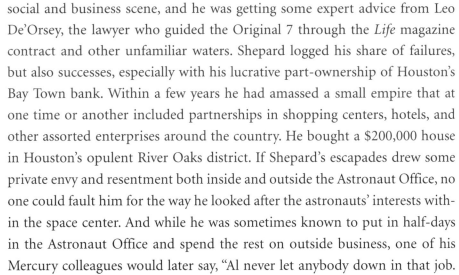

That was the beauty of Al Shepard; he could do both." On the fifth anniversary of Shepard's flight the Astronaut Office threw a party for him, and they showed a movie called *How to Succeed in Business Without Really Flying Very Much,* a slight alteration of the title of a very popular Broadway play. By that time, at age forty-two, Shepard was on the way to becoming a millionaire. ❨

But none of this mattered to Shepard compared with making another spaceflight. By the spring of 1968, his astronaut career was slipping through his fingers. His condition had worsened; he still had attacks of vertigo, and he was nearly deaf in his left ear. But Tom Stafford had heard about a Los Angeles ear surgeon named William House who had devised a delicate, risky operation to help Ménière's sufferers. The procedure involved implanting a small silicone tube in the ear to drain the excess fluid from the semicircular canals, through the mastoid bone, to the top of the spinal column. Shepard went to see House, who warned that it

Long grounded by an inner-ear disorder, Shepard gazes skyward to follow the launch of the Gemini 11 in September 1966. By this time, with his spaceflight career in limbo, Shepard was becoming wealthy from a variety of business ventures, including real estate and banking.

would take some time to know whether the operation would be a success; he gave no guarantees. In the summer of 1968 Shepard secretly checked into a Los Angeles hospital as Victor Poulos (a name chosen by the admissions nurse, who was Greek). Shepard had ruled out the surgery until he had nothing left to lose; that time had come.

●◐○○○○◐◉

In the summer of 1968, Stuart Roosa didn't have the luxury of worrying whether he would ever make a second spaceflight; he was still waiting for his first, and things didn't look good. Like his colleagues in the Original 19, he'd been immersed in rookie's labors for the better part of two years. A thin, red-headed Oklahoma native, Roosa was thirty-five, but he seemed not to have aged a day since he was a smoke jumper for the U.S. Forest Service in Oregon, fifteen years earlier. On their first date, he had told his future wife,

Roosa was thirty-five, but he seemed not to have aged a day since he was a smoke jumper for the U.S. Forest Service in Oregon, fifteen years earlier.

Joan, that he was going to be an engineering test pilot. But if there had been no Edwards and no NASA, Stu Roosa would probably have gone to Vietnam, for he had spent much of his adult life preparing to fly in combat. As a newly minted air force fighter pilot he'd spent a year in gunnery training. Then, in 1955, he was assigned to the 510 fighter-bomber squadron in Langley, Virginia. The 510's official designation was special weapons; in plain English that meant nuclear bombs. The 510 trained for the unthinkable, a full-scale nuclear war with the Soviet Union. The plan, which remained secret for years, called for the pilots to deploy out of bases in West Germany. Flying their F-100s over Russia, they would drop down to 50 or 100 feet to avoid the enemy's radar. At carefully chosen points, each man would arm his weapon and execute a series of precise maneuvers: Pull back on the stick, watch the airspeed like a hawk, and at exactly the right moment, with the jet zooming almost straight up, release the bomb, which would then follow a two-mile-long ballistic arc to the target. Meanwhile, the pilots would head in the other direction as fast as their jets could take them, to be safely out of range of the ensuing destruction. The planes did not carry enough fuel to reach their targets inside Russia and make it back. The pilots planned to eject, hundreds of miles from the base, and then, making use of special training in escape and evasion, walk the rest of the way.

Decades after the fact, it may be difficult to understand why Stu Roosa, like the other men in the 510, was not only prepared to carry out this mission, but half-wished he would get a chance to. But in 1955, at the height of the Cold War, the threat of nuclear war was a grim fact of life. Communism was the enemy, and Roosa's desire to defend his country from it was exceeded only by his love of flying. He would never forget the first time he was cleared to take off on his own as an air force cadet. In the open canopy, with the sun on his face and the roar of the engine filling his senses, he began to sing the air force song: *Off we go, into the wild blue yonder.* . . . If that sounded like *Life* magazine's image of an astronaut, then in fact, Roosa was it, more than many of his colleagues; he was a straitlaced, conservative family man with a soldier's devotion to his country. And when he became a member of the Original 19 in 1966, he was ready to dedicate his energies to the "peaceful war" that was Project Apollo.

No sooner had he arrived at NASA than he attended the party the astronauts threw for the fifth anniversary of Al Shepard's Mercury flight. He looked around and felt awed in spite of himself. He turned to his wife, Joan, and said, "Do you know that the first person on the moon is in this room tonight?"

Air force cadet Stu Roosa climbs aboard a T-6 trainer at Moultrie Air Force Base, Georgia. After winning his wings in March 1955, Roosa embarked on a flying career that would include service in a fighter-bomber squadron armed with nuclear weapons.

Roosa had no illusions that that person might be himself or any of the Original 19. He was just one more in a sea of faces at the Monday-morning pilots' meetings. He and his friend Charlie Duke were assigned to cover the Saturn boosters; to them it felt like a backwater. But as the time for the first Apollo missions neared, the astronauts began to show more interest in Roosa's work. And then, one day in the fall of 1968, he was at North American when Al Shepard called him to say, "We want you to be on the support crew for Apollo 9." Then he added, "Just be patient. I've got something in the works." Shepard's tone was so matter-of-fact that the comment barely registered; Roosa was just glad that he was finally moving out of the pack.

Communism was the enemy, and Roosa's desire to defend his country from it was exceeded only by his love of flying.

Roosa's assignment was to help coordinate the activities in mission control, and he logged months of simulations with the flight controllers. When Apollo 9 flew, Roosa was more involved than anyone expected. After Rusty Schweickart suffered motion sickness, Roosa was part of the scramble to reorganize the schedule and save as many of the mission objectives as possible. He knew more about the flight plan than any astronaut who was still on earth, and he was at the Capcom's mike almost every minute that McDivitt's crew was awake. When the other controllers left at the end of their eight-hour shifts, Roosa just stayed where he was. After it was all over, Deke Slayton told him, "You were more on top of it than the flight directors were." Just as importantly, Chris Kraft took notice—and Roosa had always felt Kraft had a great deal of influence on an astronaut's career. But Roosa still had no sense of where he stood, and he didn't know what to make of it when one day Deke asked him, "Has Al talked to you about your new assignment?"

Al Shepard had always been a mysterious, intimidating figure to Roosa. Going to Al Shepard's office was as nerve-wracking as going to see the General —Roosa always made sure his thoughts were organized and his hair was combed. But around the fall of 1968, Roosa had begun to notice some subtle changes in Big Al. At the pilots' meetings, Shepard showed an unusual interest in the tech-talk—not just whether a man was on top of his work, but

Alan Shepard's secretary, Gay Alford, posted this barometer of her boss's changeable disposition near her desk, changing the picture to let the other astronauts know what to expect. This grimacing orangutan might persuade an astronaut to postpone an encounter with the Chief Astronaut.

whether they are going to get a smile or not. Here, he flashes his good morning look.—UPI Telephoto

MOOD OF THE DAY

the details of what he was saying—some new change in the spacecraft or a mission plan. For the first time since anybody could remember, he started showing up at the gym. Every now and then Roosa would see him in mission control, talking to flight controllers. Roosa had no idea what was going on, until one day in April, when he was summoned to Shepard's office, along with Ed Mitchell. When they got there, Shepard was grinning broadly. "If you guys don't mind flying with an old retread, we're the prime crew of Apollo 13." ⟨

Roosa couldn't believe what he was hearing—no rookie had ever gone straight onto an Apollo prime crew without serving as a backup. He said to Shepard, "Did you say prime?"

Shepard flashed him a steely glance. "I said prime."

The surgery had worked. By the spring of 1969, the doctors had pronounced Shepard whole. He was back on flight status, and ready to take any mission he could get. He would have been happy to fly in earth orbit on Apollo Applications, if that's where Deke wanted him.

At a small press conference, a reporter asked Shepard who would take his place helping Slayton with the crew assignments, and Shepard said he guessed Deke would have to do it all by himself, and then, barely containing a laugh, he added, "Of course, my ol' buddy Deke, good ol' Deke . . ." and the place broke up. Someone asked, "Will you campaign?" Shepard replied, "You're not supposed to." But then, he didn't have to. His last act as Slayton's coauthor on crew selections was something no other astronaut could have pulled off: he recommended himself for the very next flight available. In the spring of 1969, that was Apollo 13. Slayton obliged him.

Word got around the Astronaut Office that Alan Shepard was now the commander of Apollo 13, and the reactions were mixed. Behind closed doors, more than one of the pilots expressed resentment that Shepard could walk right onto a prime crew without doing his turn as a backup. In reality, Shepard's ego ruled that out; nor was there any danger he might have to accept middle or right seat. Others realized that it was pointless to say that Al Shepard was cutting in line; he had been in line from the day he arrived. And as far as Deke Slayton was concerned, if Shepard wasn't qualified, no one was. ☾

Only one other member of the Original 7 was still on flight status; for him, the news of Shepard's assignment was grim. For years now, Gordo Cooper's star had steadily fallen. No one would argue that he had flown a spectacular Mercury mission—by some accounts, the best of the series. But after that, he had begun to slack off. He had always raised casual to new heights; it was Cooper who, perched atop a fully fueled Atlas booster, waiting out a hold in the count, fell asleep. His strap-it-on-and-go attitude was legendary, but unfortunately he seemed to carry it into training. On his Gemini 5 flight the other astronauts had to goad him into the simulator. No one who witnessed that episode was surprised when his next assignment was a dead-end stint on the backup crew of Gemini 12. When Slayton gave him

Gordo Cooper *(second from left)* **shares a laugh with Pete Conrad before their Gemini 5 mission in August 1965. It would prove to be Cooper's last spaceflight; in 1969 he was passed over for command of a lunar landing mission in favor of Alan Shepard.**

another backup post for Apollo 10, some wondered whether Cooper would ever fly again.

Cooper didn't help himself by entering a twenty-four-hour road race at Daytona, while he was in training. Slayton was no happier about that than a movie studio would be about a major star doing his own stunts—too much was invested in him to subject him to unnecessary risk. When Slayton pulled him from the race, Cooper bitched to the press, "They ought to hire tiddly-winks players as astronauts." In any case, Slayton had named him to back up Apollo 10 with the idea that he'd probably be able to fill in for Tom Stafford if necessary, but he was not at all sure that he'd follow the rotation and give Cooper command of Apollo 13. When Shepard became available, a few weeks before Apollo 10 flew, Slayton's choice was easy. Cooper would later hold an angry press conference, saying that Shepard had unfairly edged him out of his mission, but the truth was that Cooper had let the moon slip beyond his reach. And so Alan Shepard would become the only one of the Original 7 to reach the moon. ☾

But not on Apollo 13. For the first time since he had begun selecting crews, Slayton's assignment was vetoed—by George Mueller, head of manned spaceflight at NASA Headquarters. Mueller insisted that after so long on the bench, Shepard needed more time to train. Slayton argued that Shepard had been in training for 13 since the spring and was making good progress. Shepard himself knew he had some catching up to do. But he'd had a head start; even while he'd been grounded, he had been learning the basics of the command module and the LM, and like every rookie astronaut he'd been grabbing simulator time whenever he could get it. He felt confident he'd be ready in time for Apollo 13. But Mueller held firm. Slayton had no choice but to swap Shepard's disappointed crew with Jim Lovell's, who had just finished backing up Apollo 11. One day in August, Slayton took Lovell aside and asked him whether he and his team could be ready for 13; and Jim Lovell—happily surprised at the offer—said yes.

●◑◖◯◯◗◐●

For Alan Shepard the events of April 1970 had special irony. In the years after the Fire, it had occurred to him that being grounded might have saved his life; he might well have been the commander of Apollo 1 instead of Gus Grissom. Now he had been saved from the ordeal of Apollo 13 by George Mueller. ☾

Shepard wasn't alone in such thoughts. Stu Roosa had been in mission control when Apollo 13 became a struggle to save the lives of three men.

Initially he had been one of those who thought there was no hope for Lovell's crew, but over the next four days he'd witnessed the most impressive recovery in NASA's history. In Houston, as the accident investigators issued their report and engineers went to work on safeguards for later missions, the flight controllers savored a new confidence. Whatever problems might come up, they knew they would handle them. They were eager to take on Apollo 14, which had been postponed from July of 1970 until the end of the year.

But in the wake of their masterful rescue—which NASA called "a successful failure"—Apollo once again faded in the national consciousness. Even during the ordeal of Apollo 13, amid global concern, the country's attention had been divided. While NASA struggled to save Lovell's crew, a professor at Duke University encountered one of his students and asked, "Do you think they'll get them back?" The student responded by talking about the American troops in Vietnam. At the beginning of May, the U.S. invasion of Cambodia intensified the campus unrest over the war. Days later, during a demonstration at Kent State University, four students were killed by young, nervous National Guardsmen. By summer, the nation's conflict over the war in Vietnam had deepened.

It was also a troubled time for NASA. The relative austerity in the space budget continued to affect the program. At the Cape, there were bumper stickers that read, "Apollo 14: One Giant Leap for Unemployment." At the end of August, faced with additional cuts, acting NASA administrator George Low, who had taken over after Paine's departure, canceled two more Apollo missions. In the Astronaut Office, morale skidded as many saw their chances to go to the moon slip away. ☾

And the astronaut image was showing signs of wear. It had been a year since Donn Eisele left his wife for another woman. For years astronauts

Ed Mitchell *(top),* Stu Roosa, and Al Shepard strike an informal pose on the ladder of a lunar module trainer. Although Shepard had made a 15-minute suborbital space-flight in 1961, space-center wags dubbed them the "all-rookie" crew.

had kept shaky marriages together because no one knew how the alternative would affect a spaceflight career. If the Eiseles' split stunned the Astronaut Office, it tore apart the community of wives, many of whom were forced to admit, at least privately, what they had carefully ignored for so many years: Their husbands were playing around. The Astronaut Wives Club, which had been the nucleus for social gatherings, was so shaken over whether to invite Eisele's new wife that it ultimately dissolved. Eisele left NASA without flying again, and some wondered whether his divorce had been the reason. But at the end of 1969, Al Worden, assigned to Apollo 15, showed that an astronaut could end his marriage and remain on a space crew. ☾

Shepard was the best commander Roosa could have asked for; he nicknamed him "Fearless Leader."

For the most part, Stu Roosa had little time to think about these events, or anything else besides Apollo 14. It was true that he had pilot friends in Vietnam, and as always, he felt a little guilty about not flying alongside them. When one of them came home and expressed anger over the way the war was being handled and bitterness at the dissent at home, Roosa shared his distress. But he was more upset over what was happening to Apollo; he could not believe the nation was abandoning the most magnificent exploration in its history. The impact on Roosa was very real. He had known that as command module pilot of Apollo 14, assuming he stayed in the rotation, he would eventually command Apollo 20—but his hopes of landing on the moon had died in January 1970, when Apollo 20 was canceled.

But Roosa's personal concerns paled beside the demands of his mission. In addition to the normal training workload, he had to look after the modifications to the command and service modules. And Shepard had done something that impressed Roosa greatly: he'd let Roosa handle it. He didn't micro-manage; he wasn't afraid to delegate. He had faith in his crew, and Roosa had every faith in Shepard. The man was *very* sharp, a quick study, and in Roosa's opinion, the most competent astronaut in the office. He was also probably the most complicated man Roosa had ever known. But one thing was certain: this wasn't the same Big Al who had lorded over the Astronaut

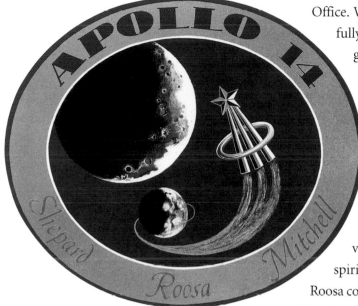

The mission emblem for Apollo 14 shows an astronaut pin flying from the earth to the moon. Shepard designed the emblem as a gesture of esprit de corps toward the entire Astronaut Office.

Office. When he came on flight status, he stepped down gracefully to the trench work of a mission commander. He was genuinely pleasant to work for. Roosa introduced him to his parents and they were charmed by him. Once Shepard accepted a person into his inner circle he could be surprisingly open. Of course, he could still turn to ice without warning—which is to say, he was still Al Shepard—but he never directed that anger at his crew. And when it came time to make up a mission patch, Shepard sketched a design that showed an astronaut pin flying from the earth to the moon, to convey that the entire Astronaut Office was going along, in spirit. In his own way, Shepard was the best commander Roosa could have asked for; he nicknamed him "Fearless Leader." ☾

Ed Mitchell, the man who would land on the moon with Shepard, had a doctorate from MIT and test pilot credentials from the space school at Edwards. In the Astronaut Office, it was his intellectual bent that set him apart from some of the other pilots, along with a certain hard edge. By his own admission, he had an impatient streak, and when angered he was capable of outbursts of temper. But when Mitchell spoke, it was with the soft voice of a midnight FM-radio announcer. Roosa shared an office with him and used to wonder how anybody at the other end of a phone conversation with him could understand what he was saying. But Roosa respected him as a professional. What really mattered was that Mitchell do his job in the command module, and that he did very well. And from what Roosa could tell, Mitchell—who matched Fred Haise in his knowledge of the LM—seemed to be carrying the load for Shepard with the lander's systems, letting his commander concentrate on the piloting. ☾

Shepard, Roosa, and Mitchell trained for nineteen months, longer than any crew before them, but even now Roosa did not really think of them as friends. Al Shepard wasn't Pete Conrad, and you wouldn't see the crew of Apollo 14 going out for a beer together at the end of the day. And Roosa couldn't have cared less. What really mattered—and what they shared with every crew before them—was a burning desire for what Roosa called a full-up mission, that is, every objective on the flight plan accomplished. By January 1971, NASA was ready to send Shepard's crew to the moon to complete the mission Lovell's team had been denied. Apollo 14 would be the first flight devoted solely to the scientific exploration of the moon. Using the pinpoint landing technique proven on Apollo 12, Shepard and Mitchell would make

A thousand lights glitter as crews work through the night to ready Apollo 14 for its journey to the moon two weeks before launch on January 31, 1971. At this point in pre-launch activities, technicians are moving the huge framework at near right, called the Mobile Service Structure, away from the rocket after a successful practice countdown.

Lighting up an overcast afternoon, the first-stage engines of Apollo 14's Saturn V booster build thrust seconds before liftoff on January 31, 1971. Dick Gordon, veteran of the lightning strike on Apollo 12, was watching with astronaut families, hoping the rocket would avoid rain clouds.

the first landing in the lunar uplands, at a place called Fra Mauro. While they made two moonwalks, including a climb to the rim of the 1,100-foot-diameter Cone crater, Roosa would survey the moon from orbit with a special high-resolution camera called the Hycon. Roosa sensed that this time, a full-up mission had special importance: NASA couldn't afford another failure. In 1961 Shepard had been the man of the hour; as far as Roosa was concerned, there wasn't anyone better to entrust with the future of the space program than Alan Shepard.

There were some people who wondered why, at age forty-seven, having acquired fame, wealth, and status as an American hero, Alan Shepard would risk his life to go to the moon. The passage of ten years had deepened the lines around his mouth and his blue, slightly bugged eyes, but the toothy grin was the same as it had been on the cover of *Life* magazine in 1961. And though the military crew cut was gone—his hair now fell partway across his forehead—his reasons for wanting to fly in space had not changed. In Shepard's mind, to say that he was brave, to call him a hero, was hopelessly incomplete: he was a supreme test pilot, and nothing mattered more to him than getting a chance to prove it. A decade ago, lying atop that gleaming Redstone rocket, he had watched as technicians sealed him inside *Freedom 7,* and had steeled himself: *Okay, buster, you volunteered for this thing. . . .* On January 31, 1971, Shepard led his crew to Pad 39-A, where a booster a hundred times more powerful was waiting for them. There was more thrust in the command module's escape rocket than there had been in the Redstone. His space-flight experience was surpassed the moment Apollo 14 reached earth orbit. Two hours and twenty minutes later, when the Saturn's third stage ignited, Shepard was heading for the goal he'd kept in sight for nearly a decade—not the moon, but the chance to show what he was made of. Years later Shepard would say that he always believed in pushing out the frontier; it was nice that the moon happened to get in the way. ☾

II: TO THE PROMISED LAND

SUNDAY, JANUARY 31, 1971
7:41 P.M., HOUSTON TIME
4 HOURS, 38 MINUTES
MISSION ELAPSED TIME

Apollo 14 almost ended before it could begin. As launch time neared Shepard's crew had looked ahead to their mission with an almost arrogant confidence: With the intensive scrutiny after Apollo 13, nothing should interfere with their mission. But now, only two hours after a perfect Translunar Injection burn, Shepard, Roosa, and Mitchell found themselves waiting for word from Houston, wondering whether they could continue. The trouble had begun shortly after Stu Roosa cut loose from the Saturn third stage and prepared to dock the command module *Kitty Hawk* with the lander *Antares*. In the past nineteen months he had simulated the maneuver so many times that he used less fuel than any command module pilot before him. Now he was doing it for real, and he wanted that fuel record. Floating above the center seat, Shepard peered through the hatch window as if he were hanging from a ledge by his fingers. He could see *Antares* dead ahead, and he told Roosa, "Gonna break the record, man."

Slowly, Roosa closed in. At last the two ships met, but a moment later they drifted apart. Roosa was mystified. He decided he hadn't come in fast

---

No one had to say what they all knew: if they couldn't link up with Antares, the mission was over.

---

enough to trigger the docking latches. He would have to back off and come in a little faster, and he knew that would cost him more fuel. "There goes the record," he said mournfully. Soon his disappointment turned to concern as the second attempt failed. Something was wrong with the docking mechanism. Over the next hour and a half, in consultation with Houston, Roosa made two more attempts without success. No one had to say what they all knew: if they couldn't link up with *Antares,* the mission was over.

Shepard's crew wasn't about to give up. The culprit might be something as simple as a tiny piece of debris on the mechanism; if so, it should eventually disappear. But they couldn't afford to wait indefinitely; in a few hours the Saturn's third stage was going to vent its extra fuel, and when that happened they had better be at a safe distance. If Houston didn't come up with a fix, Shepard thought, they would have to take matters into their own hands—literally. After putting on helmets and gloves they would depressurize the cabin, open the command module's top hatch, and bring the probe inside; there they might be able to fix it. And if they couldn't, Shepard had another idea. After Roosa steered *Kitty Hawk* back to *Antares,* Shepard would reach through the tunnel and pull the two ships together by hand; when the ships met, he was sure the docking latches would engage automatically. His only

worry, one shared by his crew, was that the managers in the back row of mission control would not let them risk it. And five days from now, there would be another linkup in lunar orbit when Shepard and Mitchell returned from the surface. Trouble then would force the men to make an emergency space walk to the command module. They'd trained for every conceivable emergency, including that one. But would Houston accept that possibility now, when it would be far less hazardous to abort the flight? ❬

"Hey, Stu, this is Geno." It was Gene Cernan's voice from mission control. "We've got one more idea down here. . . ." Cernan told Roosa he should fire *Kitty Hawk*'s thrusters to hold the command module against the LM. As he did so, Shepard would flip the switch to retract the docking probe out of the way. If the two craft were lined up properly, Cernan said, the contact might just trigger the docking latches. One hour and 42 minutes after his first attempt, Roosa once more steered a course for *Antares.*

"About six feet out," Mitchell said. "About two feet."

"Here we go," Roosa said.

They felt the ships touch. "Okay," Roosa said, "retract."

Shepard hit the switch and waited. "Nothing happened."

"Nothing?"

"I don't know." Suddenly Shepard saw the indicators on the panel in front of him go from gray to a pattern of stripes. "I got barber pole," he called. At that moment he heard the telltale "ripple-bang" of twelve docking latches snapping shut. Shepard announced, "We got a hard dock."

An hour later the joined *Kitty Hawk* and *Antares* were drifting away from the third stage on a course for the moon, and Shepard, Roosa, and Mitchell settled down to finish up what was becoming a long first day in space. It had been a close call; it would not be their last.

MONDAY, FEBRUARY 1, 1971
7:37 A.M., HOUSTON TIME
16 HOURS, 35 MINUTES
MISSION ELAPSED TIME

In the dark, floating in *Kitty Hawk*'s left-hand seat, a tired Stu Roosa tried to settle in for the night. Like those who had come before him, Roosa found it difficult to adapt to sleeping in zero g; he wished he could lay his head on a pillow. Forty-five minutes later, still awake, Roosa noticed a light coming from underneath the right-hand couch, where Ed Mitchell was in his sleeping bag. Roosa assumed Mitchell had turned on his flashlight because he'd gotten tangled in a strap. He could not have guessed the real reason Mitchell was awake, that he was conducting his own private experiment

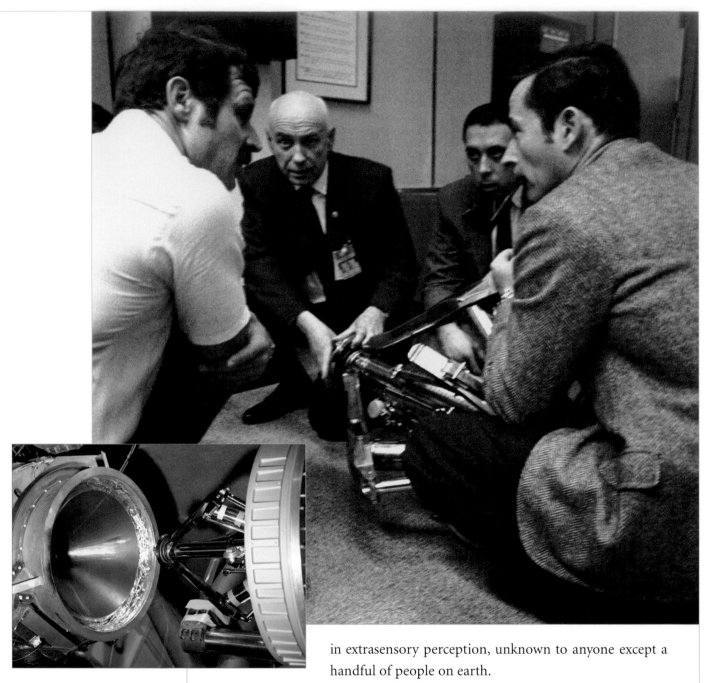

Space center director Bob Gilruth *(facing camera)* confers with a team of experts over an Apollo docking mechanism, struggling to understand a potentially mission-ending problem: The device wouldn't latch into a conical fitting called a drogue *(inset)* when Roosa tried to dock the command module with the lander.

in extrasensory perception, unknown to anyone except a handful of people on earth.

Edgar Dean Mitchell grew up as a rancher's son amid the dust storms of the Texas panhandle and eastern New Mexico. His father was a man of nature who had a great love for his animals; his mother was a fundamentalist Baptist. As a boy, sitting beside his parents in church on Sunday mornings, Mitchell was confronted by an inner conflict. The preachers in those small churches were the fire-breathing, Bible-waving kind, and the younger Mitchell found himself doubting the existence of the God they praised and the Satan they warned of. By the time he reached high school he had concluded that the Creation story was allegory, not literally true as the preachers insisted. Their stern admonitions against social and sexual transgressions, even dancing with girls, repelled him. As much as he respected his parents and the church he was raised in, as much as he sensed a great body of truth in its teachings, he could not follow that path. His strong interest in math and science only deepened his sense of irreconcilable conflict, for neither side seemed to acknowledge the other. From then on, Mitchell hungered for resolution, and he carried into adulthood a desire to understand the nature of the universe. It was still with him when he became an astronaut.

Mitchell was probably the only astronaut who missed the presence of psychologists in the space program.

When the press asked Mitchell why he wanted to go to the moon, he told them it was the logical extension of everything he'd been doing as a fighter pilot and then as a test pilot. He was also quick to mention his scientific curiosity, and in truth, the greatest lure was the chance to explore the unknown. When he arrived at NASA the Nineteen had little hope of flying on Apollo, but Mitchell had his sights on a more distant goal: to captain the first manned expedition to Mars. By 1970 it was clear that NASA would have to abandon for the time being any plans of sending people to the Red Planet, but by then Mitchell was already training for Apollo 14. And even as he looked forward to probing the secrets of Fra Mauro, he saw a chance to explore a frontier even more mysterious than the moon: the nature of consciousness.

Mitchell was probably the only astronaut who missed the presence of psychologists in the space program. NASA had engineers to handle the technology of flying to the moon, and scientists to puzzle over its geologic riddles, but no one to unlock the inner experiences of the men who had been there. About three weeks before the flight, a chance conversation inspired Mitchell to take advantage of the fact that he would become one of the few human beings to leave the planet. As some of his colleagues knew, Mitchell had long been fascinated by the study of psychic phenomena, for which neither science nor religion offered a satisfying explanation. He'd become acquainted with a couple of surgeons in Florida who shared his interest. Together they

Ed Mitchell *(above)* saw his Apollo flight as a unique opportunity for an experiment in the paranormal phenomenon of extrasensory perception. In secret, he arranged to transmit mental images to several individuals on earth—among them, a Chicago businessman named Olof Jonsson *(right)*—during sleep periods on his voyage to the moon and back. And so he did, but on a schedule, drawn up before the flight, that did not account for several hours of holds during countdown. As a result, Mitchell's collaborators at home were not tuned in during his transmission efforts, disrupting the experiment.

wondered, was it possible to transmit thoughts across a hundred thousand miles of space? In the midst of the all-consuming preparation, Mitchell told them, "Line up some people and we'll do a little experiment on the flight." ☾

And so they did. Each night of the trip to and from the moon, Mitchell planned to perform the experiment, waiting until forty-five minutes past the start of the sleep period, when he had privacy and quiet. He kept his plan a secret from NASA, knowing that the agency would be completely unreceptive to the idea. He said nothing about it to his crewmates. The test subjects had also agreed to keep quiet. And Mitchell wasn't worried about what would happen if someone found out; with all the canceled missions he was already certain that Apollo 14 would be his only spaceflight.

Now, floating in his sleeping bag, Mitchell pulled out a small clipboard bearing a table of random numbers. Each number designated one of the standard symbols used in ESP experiments: a circle, a square, a set of wavy lines, a cross, a star. Mitchell chose a number and then, with intense concentration, imagined the corresponding symbol for several seconds. He repeated the process several times, with different numbers, knowing that on earth, four men were sitting in silence, trying to see the pictures in their own minds. After several minutes of this, Mitchell put the paper away and closed his eyes. ☾

Had Roosa known of Mitchell's activities, he wouldn't have been too surprised. In Roosa's opinion, *Kitty Hawk* was carrying three people who, within the narrow military-test-pilot filter, were as different from one an-other as it was possible to be. On the circle of astronauts, they spanned a full 360 degrees. Deke Slayton always said he could put the three most divergent personalities in the Astronaut Office on the same crew and they would do just fine. As far as Roosa was concerned, the Apollo 14 crew proved him right.

Ed Mitchell tapped on the instrument panel with a flashlight as *Antares* drifted through darkness over the near side of the moon. Only ninety minutes away from Powered Descent, the mission of Apollo 14 was once again in jeopardy. He and Shepard had been checking the lander's guidance software when engineers in mission control detected that the computer was receiving an errant signal from the abort button. They guessed that a tiny ball of solder was floating around within the switch and closing a contact, and sure enough, when Mitchell tapped on the panel, the signal disappeared. For the moment it had no effect, but if it came up when Shepard and Mitchell lit the descent engine, the computer, not knowing any better, would read the mistaken signal and automatically abort the landing. Since Apollo 13, mission control had trained for every malfunction in the book—and there literally was a book, with procedures for any conceivable emergency—but not this one. ☾

Once more there was nothing to do but wait for mission control to come up with a solution. In Houston, and in Cambridge, at MIT, computer programmers began a feverish effort to work around the malfunction. They would have to tell the computer not to accept any signal from the switch at all, to "lock out" the erroneous command. But that would mean rewriting a portion of the LM computer's software—right away. At MIT a young programmer named Don Eyles went to work; he was done before Shepard and Mitchell went behind the moon for the last time before the scheduled Powered Descent. By the time *Antares* reemerged, the programmers had streamlined and improved the fix, and relayed it to Houston.

With only minutes to go before ignition, Mitchell keyed in the necessary changes, telling himself, "It's just like a simulation," a good trick to stay calm. He and Shepard had been through some horrible scenarios in the simulator; by comparison, this was almost a leisurely operation. But when the descent engine lit at 3:05 A.M., Shepard and Mitchell held their breath. The fix worked. As the engine rumbled silently on its long brake, Mitchell keyed in a few additional changes, and they kept going. Once again the experts on earth had saved the mission.

3:09 A.M.

Four minutes later, as *Antares* descended through 32,000 feet, everything still looked good. The next step was for the landing radar to lock onto the echoes of its own signals bouncing off the surface of the moon. But on the computer display, caution and warning lights glowed, signaling that the radar had still not locked on. "C'mon, radar," Mitchell said quietly, "let's have the lock-on." Still no change. "*C'mon,* radar."

"Go at five," Fred Haise radioed. Six minutes into the burn now. Velocity was good; altitude estimates by both the primary and backup computers were in agreement. But still no radar. Shepard and Mitchell knew that if the radar

In a lunar module simulator at MIT, LM software expert Don Eyles *(right)* and a colleague discuss an errant abort signal to the lunar module's computer *(inset)*. Had Eyles not solved the problem, the computer would have ended *Antares*'s descent as soon as it began.

didn't come in by 10,000 feet the mission rules specified a mandatory abort.

Now Haise told them, "We'd like you to cycle the landing radar breaker." Shepard pulled the circuit breaker out and pushed it back in again. "Okay," he radioed, "it's cycled." *Antares* was down to 22,000 feet now, and Mitchell could no longer hide his urgency: *"Come on."*

Suddenly the caution lights went out, and the radar signals began to come in. Within seconds the men could see that its data were good, and Houston confirmed that fact. Once more, with help from mission control, they had made a narrow escape. And when *Antares* cleared 8,000 feet and pitched over, Shepard and Mitchell were electrified. Cone crater was dead ahead, right where it should be. "Fat as a goose," Shepard said, and proceeded to steer *Antares* to a smooth touchdown, closer to his target than any other landing in the Apollo program. ⟨

When the dust had settled and they were on the moon to stay, Mitchell thought back to their close call, when *Antares* had been flying beautifully and only the balky radar and the mission rules had stood in their way. He asked his commander, "What would you have done?" Shepard wasn't about to say what he was thinking, that he would probably have rewritten the mission rules on

This view of Cone crater *(arrow)* rising from the Fra Mauro highlands greeted Shepard and Mitchell as *Antares* neared the moon. The picture was recorded by the onboard movie camera.

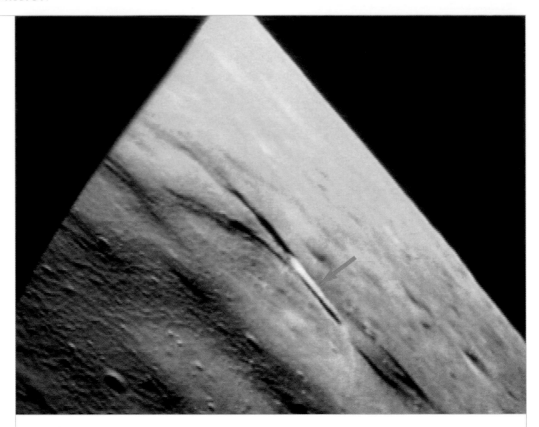

the spot. Radar or no radar, he would have continued as long as everything still looked good. To Mitchell he said, simply, "You'll never know."

8:54 A.M.

It is impossible to look up at the full moon and not notice Mare Imbrium, the huge dark splotch that forms the Man in the Moon's right eye. Down where his right cheekbone would be, the lunar module *Antares* rested on a region of hills and craters called Fra Mauro. The roughness of the place surprised Shepard and Mitchell, who had expected to see something resembling the *mare* plains visited by their predecessors. Instead it bore the scars of eons of cosmic bombardment. Shepard was filled with the realization that time here was measured in billions of years rather than the thousands of years spanned by human history. Nevertheless, just before 9 A.M., when Shepard stepped off *Antares*'s foil-covered footpad, the words he spoke were not for the timeless moon, or even for the history books, but to mark the end of his own personal odyssey:

"It's been a long way, but we're here."

Finding his balance, Shepard made his way to one side of *Antares* and cast his gaze to the east. There, beneath the solar glare, was the broad rise of

Cone crater. Tomorrow he and Mitchell would climb to its summit in search of geologic treasure. He told Houston the way would be clear; he could see that even from where he stood. Shepard realized that, finally, everything was going well. After all they had been through, he felt sure he would have his full-up mission. He took a moment to lean back so that he could look up into the black sky, and near the zenith his gaze found a small and lovely blue and white crescent. Suddenly he was overcome by the beauty of the earth, by the undeniable majesty of Fra Mauro, and by his own feeling of relief. Standing on the gray dust of this promised land, Shepard cried. For several long moments, while the checklist went unnoticed, his tears flowed in spite of himself.

III: SOLO

Stu Roosa hadn't slept well, but when *Kitty Hawk* came around from the far side of the moon on his fourteenth revolution alone, he lied about it and told Capcom Ron Evans he'd gotten six hours of good sleep. In fact he'd had a fraction of that, and none of it good, but he didn't want anyone on the ground to worry about him. And thanks to a steady flow of adrenaline, he felt well. But as he brought *Kitty Hawk*'s systems back up for another day, Roosa took a deep breath. Yesterday had been long; today would be even longer. His solo voyage was turning out to be a forty-one-hour fire drill.

Roosa was pleased at how well prepared he'd been for this whole business, so well that on the trip out—to his surprise—he did not experience awe or amazement at seeing the earth shrink to the size of a marble: after hearing five Apollo crews talk about the dwindling earth, there wasn't much impact when he saw it for himself. It wasn't until Apollo 14 was more than halfway to the moon that he sensed the immense distance; it came not by looking but listening to the long moment of silence between his call to Houston and the response. As for the conversations themselves—well, the air-to-ground transcripts from this mission were going to be just plain dull. The talk was all technical, and that didn't surprise Roosa one bit, not with this crew. There were no weather reports from space, no lighthearted exchanges with mission control. It was thirty-six hours into the mission before they said anything about the view outside, but—they weren't there to talk about the view.

Roosa had thought he knew what to expect from the moon, too, but nothing had prepared him for what he saw just before Lunar Orbit Insertion, when a thin but enormous crescent moon loomed beyond the windows. On

the tape player, just by chance, was one of Roosa's favorite hymns, "How Great Thou Art." He couldn't have asked for better background music:

> *When I in awesome wonder*
> *Consider all the works Thy hands have made*
> *I see the stars, I hear the mighty thunder*
> *Thy power throughout the universe displayed*
> *Then sings my soul, my Saviour God to Thee;*
> *How great Thou art, how great Thou art.*

Gazing down on the moon from 69 miles, Roosa wished he could take an orbit to do nothing but look at it. But his mission—the first intensive program of scientific observation in lunar orbit—was far too demanding for that. It was largely the mission Ken Mattingly had meant to carry out; like him, Roosa had spent much of the past nineteen months working with the geologists. Each of his assigned craters was a geologic world unto itself, with its own peculiar features, its own riddles. His Solo Book, crammed with

---

Gazing down on the moon from 69 miles, Roosa wished he could take an orbit to do nothing but look at it.

---

photographic tasks, was as demanding as the one Mattingly had prepared for Apollo 13. On top of all this, he had the work of a command module pilot: landmark tracking, maintaining the systems, firing the SPS engine to adjust his orbit. It was a grueling workload, and time seemed to be slipping through his fingers, mostly because the single most important piece of gear for his geologic reconnaissance, an automatic, large-format camera called the Hycon (another holdover from Apollo 13), had conked out.

The Hycon was a huge thing; the lens alone was as big as the 18-inch hatch window. It had a motorized film transport and exposure controls and its own timer. With it, Roosa had planned to take photographs that would show features less than 7 feet across. Just hours after he and Shepard and Mitchell arrived in lunar orbit, Roosa unpacked the Hycon and mounted it at the hatch window. It breezed through about 140 frames; then it began making a distressing *clack-clack-clack-clack-clack,* until he turned it off. Since then, he'd spent much of his time troubleshooting. He unplugged and recon-

As the moon's two-week-long day begins, the unrelenting sun
bathes *Antares* in early-morning light. With one footpad in a
small crater, the craft tilts noticeably to one side, but not enough
to endanger the astronauts at liftoff.

nected every cable again and again. He unmounted it and peered into the lens with a flashlight while it clacked away. Houston offered troubleshooting advice, but nothing seemed to help, and he'd already lost so much time that he was way behind schedule.

Roosa was too busy to notice his isolation—most of the time. The exceptions came after sunset. As *Kitty Hawk* drifted silently through the realm of earthlight, the cabin cooled slightly—just a few degrees at first, but that was enough that the environmental control system could not remove all the moisture in the air. Then the spacecraft went out of radio contact and into total, unyielding darkness. He liked the solitude, but he couldn't deny a feeling of loneliness. His only company were the stars that filled the sky, except where the moon blotted out even these distant companions. While Roosa worked the air turned clammy, and suddenly it seemed he could *feel* the darkness. He knew, in that moment, what it was to be utterly alone.

After nearly a half hour of this—longer than most previous missions, because the landing site was well to the west, and most of the near side was in sunlight—something remarkable happened. In a finger-snap, the cabin was flooded with sunlight, and it was such a glorious feeling of renewal, even rebirth, that Roosa said to himself, "You know, we're really not creatures of darkness." Roosa realized where he was and felt great and small at the same time. It was the moon that made him feel big: He'd made it all the way out here. And it was the sight of the earth—now, for Roosa, an object of undeniable wonder and nostalgia—that made him feel small.

Only once, near the end of the first day, did Roosa have a brief respite from his flight plan. It happened after the last experiment of the day, which called for him to take pictures of the earthlit part of the moon using high-speed film. When the time came, on his eighth solo orbit, he pointed *Kitty Hawk* straight at the moon and turned out all the cabin lights. Mounted in his rendezvous window was the Hasselblad camera. All was silent except for the soft whine of the cabin fans, which he had long ago tuned out, and the steady *click-click-click* of a timer, telling him when to snap pictures. Below, shadows of morning lengthened as *Kitty Hawk* drifted westward, making the Ocean of Storms jagged and forbidding. Every ridge, hollow and pinhole was sharp. One large crater straddled the terminator, half of its rim in brilliant sunlight and half lost in the night. Even as *Kitty Hawk* crossed into blackness, he kept firing off pictures. After a few moments his eyes adapted to the darkness and the moon reappeared, bathed in soft, blue light. He photographed the earthlit ground for several minutes, and then the experiment was over, and his flight plan for the day was finished. No more experiments, just grab a

The command ship *Kitty Hawk,* named for
the birthplace of powered flight, floats in
lunar orbit. Pilot Stu Roosa's efforts to make
Kitty Hawk an orbiting research station
were partially frustrated when an onboard
high-resolution camera stopped working.

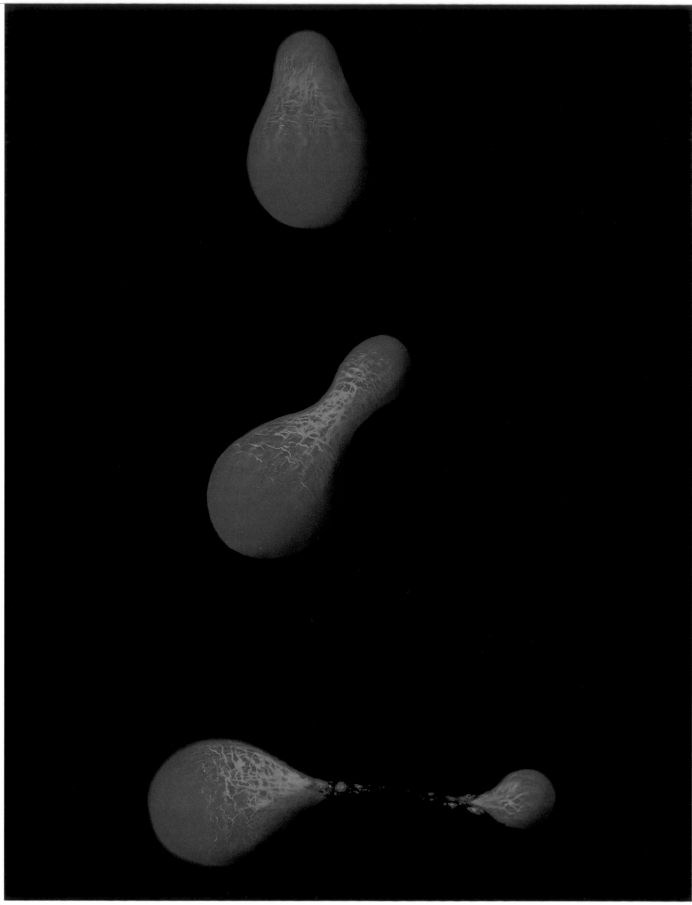

clever enough to decipher it. They took a giant step when geochemists as-
sayed radioactive isotopes in the Apollo 11 rocks and found that the samples
were 3.65 billion years old. That age, which dated the epoch when the Tranquillity lavas erupted, was the first hard nail of truth on which the geologists
could hang their timeline of lunar history. Contrary to what Harold Urey believed, the *maria* were not primordial. Geochemists had already established
from meteorites that the birth of the earth and moon had happened 4.6 billion years ago, nearly a full eon before the Tranquillity lavas poured forth.
But some hot-mooners had to face the surprising truth that the moon's volcanic activity took place far earlier than they had thought.

The geologists had barely had time to probe the Apollo 11 rocks when
Apollo 12's haul arrived. The *mare* basalts from the Ocean of Storms were a
good 400 million years younger than those from Tranquillity Base, confirm-

---

*The geologists had barely had time
to probe the Apollo 11 rocks when Apollo 12's
haul arrived.*

---

ing the geologists' belief that the *maria* did not all form at the same time.
That was consistent with geologists' theories about an era of *mare* volcanism
that may have spanned a few billion years. The Apollo 12 samples also differed in composition from the Apollo 11 rocks and from one another. That
implied that the source of the *mare* basalts, the lunar mantle, must also vary
in composition from one part of the moon to another.

Meanwhile, the dust of the moon was telling its own story. It bore the
tracks of cosmic rays and was laden with subatomic particles from the sun
for the astronomers and solar physicists to study. For the geologists, it
sparkled with tiny beads of glass. Some of these, the geologists deduced, had
formed by volcanic eruptions, possibly from deep within the moon. Others
had likely sprayed out as molten droplets from the tremendous heat of meteorite impacts. In all, the lunar samples from these two landings would have
been enough to keep scientists busy for years.

But they were not enough. A glance at the full moon reveals that the
maria are surrounded by bright-colored highlands, otherwise known as the
terrae, whose composition was still almost totally unknown. The relatively

smooth *maria* had been the safest choice for the first landings, but after the success of Apollo 12's pinpoint landing, NASA was ready to send astronauts to the more difficult terrain of the lunar highlands. There, at a place called Fra Mauro, Shepard and Mitchell were ending an uneasy night in the lunar module *Antares*.

FRIDAY, FEBRUARY 5
8:02 P.M., HOUSTON TIME
5 DAYS, 5 HOURS,
39 MINUTES MISSION
ELAPSED TIME

"Are you awake?" whispered Shepard.

In the bottom hammock Mitchell answered in a hushed voice, "Hell, yes, I'm awake."

"Did you hear that?"

"Hell, yes, I heard that."

An unfamiliar sound had jolted the men out of a fitful slumber. It had been hard enough to fall asleep wearing space suits, but they were even more uncomfortable because Shepard had put *Antares* down with one footpad in a small crater, and the whole craft was tilted noticeably to one side. Now, as they lay in the darkened cabin, the strange noise made them wonder whether the lander was resting on solid ground. Shepard whispered, "You don't suppose this damn thing is tipping over?"

After a momentary realization—Why were they *whispering?*—the two men scrambled out of their hammocks, Shepard practically falling onto Mitchell in the process, and raised the window shades—and realized that *Antares* was still firmly perched on the rolling hills of Fra Mauro. For a few more hours they tried to sleep, without success. It didn't matter; they were so anxious to begin the second moonwalk that they felt no exhaustion. ☾

Just past 2 A.M. on Saturday, February 6, Shepard and Mitchell emerged from *Antares*. Ahead lay the climb up Cone crater and a search for geologic treasure. They prepared for the traverse by loading supplies onto a two-wheeled cart called the MET (for modular equipment transporter), which carried geology tools, sample bags, magazines of film, and other gear. It also carried a message from Gene Cernan and the backup crew. Before the flight, Cernan's crew had devised a mocking version of the Apollo 14 mission patch featuring the Road Runner and Wile E. Coyote cartoon characters. On it, the Coyote, representing Shepard's crew, reaches the moon only to find the Road Runner—Cernan's crew—is already there, waving a "First Team" banner. Along the border was printed the Road Runner's trademark "Beep Beep." The *real* message was unfit for publication: *Watch your ass—we're right behind*

A wizened Wile E. Coyote arrives at the moon to find that his cartoon nemesis Road Runner has beaten him to the goal. Apollo 14 backup astronauts Gene Cernan, Ron Evans, and Joe Engle created this satire of the mission insignia, then they salted the emblem throughout the command module and the LM.

you. During the flight Shepard's crew discovered Road Runner patches in every notebook and storage locker in their two spacecraft. Even on the lunar surface, they couldn't escape: There, on the MET, was another "Beep Beep." ☾

At 2:51 A.M. the men left *Antares* and headed to the east, where Cone crater rose into the glare of the morning sun. While Shepard pulled the MET, Mitchell studied a photo-map. Their route would take them past a number of large craters, then onto Cone's flank. Reaching the upper slopes, they would head northeast along a broad ridge, following it right to the crater's edge. But almost immediately, he and Shepard had trouble spotting the craters they used as checkpoints. The place was a sea of hummocks, like sand dunes, with depressions in between. Some of them were 10 or 15 feet deep. Under the brilliant sun and black sky it was like an alien, rock-strewn Sahara—and just as difficult to navigate. He found himself looking at the map thinking, "That next crater ought to be 100 meters away," but it was nowhere in sight. Even a large crater could be so well hidden from his view that he wouldn't spot it until he was right next to it.

With some care he and Shepard managed to locate their first sampling stop, where they collected a handful of rocks, snapped photographs, and took

Even a large crater could be so well hidden from his view that he wouldn't spot it until he was right next to it.

readings with a portable magnetometer. After more walking and deliberating they found their second stop. They lingered there just five minutes, long enough for Shepard to pick up a single rock as a grab sample. Then, as Mitchell took his turn pulling the MET, he gave the rallying cry: "To the top of Cone crater."

In Houston, Mitchell's words were heard by Capcom Fred Haise, who had once planned to make this climb with Jim Lovell. Now he served as Shepard and Mitchell's link with mission control, and with a back room full of eager geologists. As he listened, Haise followed along on his own photo-

Shepard holds a core sample he obtained during the walk to Cone crater. To the left is the modular equipment transporter (MET), nicknamed the rickshaw.

map; he could also check the men's position by glancing at one of the big screens at the front of the control room. On the board, Haise saw that the men should be near the sloping side of Cone crater. Then came Mitchell's voice: "We're starting uphill. . . ."

The ground on Cone's flank was firmer, but there were more rocks, and Mitchell was forced to slow down as he threaded a winding course among the craters. Every time the MET's wheels hit a rock it lurched upward in slow motion, and he worried it would tip over and scatter equipment and samples across the landscape. Shepard finally grabbed the back of the cart and the two men carried it, while Shepard jokingly muttered "Left, right, left, right," like a foot soldier.

From earth, Haise radioed, "There are two guys sitting next to me who kinda figured you'd end up carrying it up." He didn't have to explain what he meant. Gene Cernan and his lunar module pilot, Joe Engle, had bet Shepard and Mitchell a case of scotch that they wouldn't make it to the top of Cone as long as they had to drag the MET with them. But they were determined. The view into the 1,100-foot-wide pit would be spectacular. Furthermore, the scientists had told them, the deepest rocks blasted out of Cone would lie at the rim itself. With or without the MET, they would get there. Anything less, in Mitchell's mind, would be less than a full-up mission. Just ahead, the ground sloped upward in what was surely the last rise before the summit. But the climb was far more tiring than they expected. The stiffness of their pressurized suits fought every step. ☾

As Shepard and Mitchell took a much needed rest they stole a moment to look behind them at the bright, undulating plains of Fra Mauro. Tracks from the MET's two small rubber tires stretched like shiny ribbons down the hillside, into the broad valley where *Antares* rested like a tiny scale model. Already the men were more than twice as far from their LM as any previous moonwalkers. Like their predecessors, Shepard and Mitchell found that the lack of familiar landmarks and the unreal clarity of the scene made it almost impossible to judge distances, and that only made navigating more difficult. But the climb was almost over now. From this high place, Shepard and Mitchell savored the anticipation of victory.

One thing puzzled Mitchell—if they were nearly at the rim of the crater, it didn't look anything like he expected. On one of the field trips with the geologists they'd visited a nuclear explosion crater in Nevada, and that thing had boulders the size of small cars around its rim. But there were no such rocks here. Seconds later, as he and Shepard reached the top of the rise, he knew why.

"We haven't reached the rim yet," Shepard said, his voice betraying little of his surprise. All they had done was climb over a ridge; out ahead, Cone's flank rose into the distance. Suddenly Mitchell wasn't at all sure where they were. He told Haise, "Our positions are all in doubt."

Down the hall from mission control in the Science Operations Room, a thirty-nine-year-old geologist named Gordon Swann listened to the transmissions from Fra Mauro. Since the early 1960s, Swann had brought his considerable energy and geologic expertise to the work of planning lunar exploration for the U.S. Geological Survey. Raised in western Colorado, Swann had the humor, sensitivity and political savvy to be an ideal leader of the Apollo 14 field geology investigation. Swann had helped devise the traverse Shepard and Mitchell were now struggling to complete. Over the years he had gotten to know most of the astronauts on geology trips, and had made some good friends. But he never managed to get close to Shepard and Mitchell. Before the flight, Shepard had told him, "I guess you realize rocks

In mission control, Cernan *(left)* and Engle listen as Shepard and Mitchell struggle to climb the sloping flank of Cone crater. The pair had bet a case of scotch that the two moonwalkers would not reach the crater's summit while pulling their wheeled tool carrier.

and geology aren't too big with me, but I'll try and do a good job for you."

"I can understand that," Swann had answered. "I'm not too big on aeronautical engineering."

Shepard replied, "I guess we understand each other."

Now that he was on the moon, Shepard seemed to be confirming his promise. He was making his own geologic observations; a little earlier he'd even corrected Mitchell's description of a glass-splattered rock. Swann wasn't surprised, though, that he and Mitchell were having trouble navigating. He knew well from Conrad and Bean's experience that it was one thing to see a feature on an orbital photo and another to recognize it standing on the moon, with no obvious topographic clues. The solution was training. Swann

Rickshaw in tow, Shepard climbs Cone crater's flank. At this point, he and Mitchell believed that the crater's summit lay just beyond the horizon. In reality, their goal was still a quarter-mile away.

had offered to brief Shepard and Mitchell on how to spot landmarks, and they had invited him to the Cape a few weeks before launch. But when Swann met with them in the crew quarters they seemed unconcerned. "We'll have the maps," they told him. "And you guys will be in the back room, telling us where to go." Swann did his best to get his message across, but the truth was that he and the other geologists couldn't do as well at navigating, from a quarter-million miles away, as the men who were on the moon. And now, no one in the back room knew exactly where Shepard and Mitchell were.

In his headset, Swann could hear the sound of heavy breathing; the climb was taking its toll. A few minutes ago Shepard's heart rate had hit 150, prompting a flight surgeon in mission control to request that they stop for a

rest. One of the doctors called back to the Science Support Room and asked, "How important is it to get to the top of Cone crater?"

The question was fielded to Swann, but he didn't want to answer. Cone crater was a natural excavation into the Fra Mauro, and the closer Shepard and Mitchell got to the rim, the deeper the source of the rocks they would pick up. The deepest rocks, and perhaps the most important, would lie near the crater's edge. Getting to the rim was desirable, but not paramount—but Swann didn't want to say that. If he downplayed the rim, he feared, the doctors would call Shepard and Mitchell back. If he recommended to keep going, he knew the managers in the back row of mission control might veto the request. Swann hoped the men would push on as long as they could—and as long as mission control would let them. So he filibustered. ☾

Even as he spoke, Swann listened to the voices from the moon. Shepard was arguing to spend more time on collecting samples. He had spotted some boulders up ahead which he felt sure were deep ejecta from the crater. He wanted to sample them and then turn back.

Swann could hear that Mitchell wasn't happy about that idea. "Oh, let's give it a whirl! We can't stop without looking into Cone crater," he told Shepard. "We've lost everything if we don't get there." Swann could understand the younger man's frustration. No one had ever visited a 370-yard lunar crater, and after coming all this way it was only natural that Mitchell wanted to look into it. But Swann doubted he would see anything of scientific

In the Science Support Room, a geologist notes Shepard and Mitchell's progress on a photomap. Cone crater is the large crater at the top of the picture.

importance; the unmanned orbiter photos had showed no signs of exposed layers or other features.

Just then, Fred Haise called back to ask for a verdict, and everyone in the room agreed that Shepard and Mitchell had come close enough. Swann heard Haise radio, "The word from the back room is they'd like you to consider where you are the edge of Cone crater."

Mitchell answered, "Think you're finks." But Haise responded with a bit of leeway. If they thought they could reach the rim soon, it was their decision. Mitchell wasn't giving up. And mission control was extending the moonwalk by half an hour. Swann smiled when he heard Shepard say, "We'll press on a little farther, Houston. And keep your eye on the time."

Shepard and Mitchell came to another rise and stopped to rest again. All around them, the ground was littered with rocks that must have rained out of the sky like artillery fire after the impact that formed Cone crater. Two hundred and thirty feet below, the valley was awash in sunlight. Mitchell was sure that somewhere close by, among the rocks, was the crater.

"Deke says he'll cover the bet if you'll drop the MET," Haise said. But Mitchell rejected that idea; they would need their tools when they reached the rim. "The MET's not slowing us down," he said gamely. "It's just a question of time. We'll get there."

Following his commander, Mitchell watched him pull the tool cart. He realized Shepard wasn't heading in the right direction.

"Al? Head left. It's right up there."

"Yeah. I'm going there."

They pushed onward, dodging boulders. Both men were breathing hard. Again they stopped.

"We're right in the middle of the boulder field on the west rim," Shepard radioed. "We haven't quite reached the rim yet."

When Mitchell heard those words he realized Shepard thought they were farther to the north than they really were. No wonder he'd been heading toward the rise straight ahead; he thought they were just west of the crater. Mitchell was sure they were south of it. He pulled the map from its holder on the MET. Yes, he could see where they were now. If they headed north—off to the left—they'd reach the rim. He went over to Shepard and showed him the map.

"Look," he said, his words interrupted by heavy breathing. "Let me show you something. . . . We're down *here*. We've got to go *there*."

Mitchell pulled. The MET caught a boulder and almost tipped over, but he saved it. Now the grade flattened out; they'd reached the point of maximum elevation. But a minute and a half later the rim was still nowhere in sight. Mitchell realized that once more, the moon had fooled them. Intently, he studied the map. "This big boulder, Al, that stands out bigger than anything else—we oughta be able to *see* it." They were so close. If only they had a little more time. But that was something they had just run out of.

"Okay, Ed and Al." Haise's voice was matter-of-fact, but his words meant the quest was over. They'd eaten into the half-hour extension. They couldn't take any more time looking; they had to start sampling.

Shepard and Mitchell gathered some rocks from the place where they had stopped, and then they headed down the slope, to the northwest, toward the strange, white rocks Shepard had noticed earlier. The geologists had said the rocks of the lunar highlands would be different, and these surely were,

Amid a rock field high on Cone crater's flank, Mitchell searches for a path to the crater's rim, not knowing that it lies less than 1,000 feet ahead. The moonwalkers advanced no farther, thus losing their bet with Cernan and Engle.

unlike any of the samples they'd seen in the LRL. Wielding a geology hammer, Mitchell approached the boulders, which were 4 or 5 feet high. Streaks of gray and brilliant white ran through them like pulled taffy. At the first strike an outer surface crumbled away. Mitchell hit the rock again, harder, and knocked off a chip about the size of an egg. Clumsy in his gloves, he struggled to place it in a small Teflon sample bag, rolled up the top, and placed it in the rock bag on the MET. Already it was time to leave.

On the way down Shepard and Mitchell made up for some of the time lost in the climb. Now they could really fly, bounding downhill in giant, slow-motion leaps, sailing over rocks as they went. They had no trouble navigating; fortunately their last station stop had been near a large boulder. They spotted the big rock while they were still up on the slope and kept it in sight all the way down.

Three hours after they had left *Antares* Shepard and Mitchell returned with their precious cargo of rocks and film. By now they were both very tired, but Shepard had something to take care of before the moonwalk ended, a

little "gotcha" for the TV audience. He'd dreamed it up one day during training, when Bob Hope came for a tour of the space center. Shepard and Deke Slayton were escorting Hope, and they took him over to the one-sixth g rig they used to simulate walking on the moon. Hope carried a golf club with him everywhere he went as if it were a pacifier, and he refused to let go of it even when they had him on the rig. There he was, bouncing around with that golf club in his hand, and that's when Shepard, also an avid golfer, said to himself, *There has to be a way. . . .*

Now Shepard stood before the TV camera. "In my left hand," Shepard announced, "I have a little white pellet that's familiar to millions of Americans . . ." In his right hand, he held the handle from the contingency sample collector, now slightly modified: it had a genuine six-iron at the end of it. His pressurized suit was so stiff that he had to swing his makeshift club one-handed. His first swing missed, and on the second he shanked the ball. Dropping another ball to the dust, he swung once more and made contact, and as the ball sailed away into the black sky, arcing over the craters in slow motion, Shepard announced, "Miles and miles and miles!"

SATURDAY, FEBRUARY 6
7:53 P.M., HOUSTON TIME
6 DAYS, 4 HOURS,
50 MINUTES MISSION
ELAPSED TIME

The cratered sphere filled the window, hanging in blackness. *Kitty Hawk* was climbing away from the moon on the Great Elevator Ride. Ed Mitchell felt a weariness that was not just physical exhaustion but a longing so profound as to be felt in the body. He did not want this moment to end. It wasn't merely the view that was so powerful, it was the *idea:* He and Shepard had just been down there. They had walked on another planet. Mitchell took a long, last look; he knew he would never return. ☾

By the next day the three men were still recovering from the grueling day before. Now the quiet time of the mission had arrived, three days with little to do but keep the spacecraft operating and catch up on sleep. The pressure was off. Later Mitchell would look back and realize he was in the ideal condition for what was about to happen to him.

Beyond the command module's windows, a cloud-covered, crescent earth cast its light in the blackness. By now it was a familiar sight, but every once in a while Mitchell stopped to look at it. Gradually, as he worked and glanced at the bright crescent, he was filled with a quiet euphoria, great tranquillity, and an overpowering sense of *understanding*. It was as if he had suddenly begun to hear a new language, one being spoken by the universe itself.

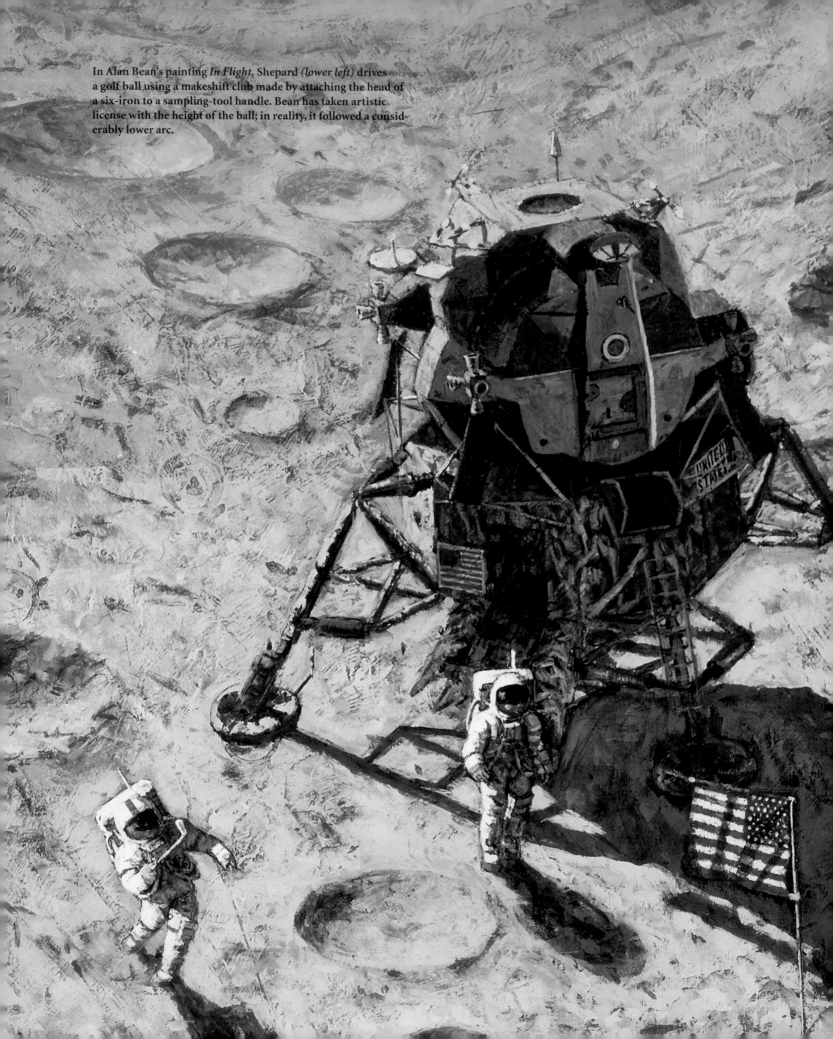

In Alan Bean's painting *In Flight*, Shepard *(lower left)* drives a golf ball using a makeshift club made by attaching the head of a six-iron to a sampling-tool handle. Bean has taken artistic license with the height of the ball; in reality, it followed a considerably lower arc.

No longer did the earth or anything in the universe seem to be random. There was a sense of order, of worlds and stars and galaxies moving in harmony. In one moment he was a detached observer; in the next he could see that he was definitely a part of it all.

As he worked within the command module he had a sense of being outside himself, as if someone else's hands—way down there—were turning knobs, flipping switches. He found himself glancing over at Shepard and Roosa, looking for some glimmer in their eyes, some sign that they were sharing any part of this awakening, but they seemed the same as they had always been. He said nothing.

Several times during the trip home, the feeling returned, triggered each time by the sight of his home planet. Mitchell knew he had been enlightened, but in a way he did not understand, and with an impact that even now, he did not fully sense. In time, it would overshadow even walking on the moon. In it, Mitchell would try to find the seeds of resolution he had longed for all his life. He did not guess, as *Kitty Hawk* coasted toward earth, that Stu Roosa, and even Al Shepard had tasted their own moments of personal discovery. In ways they would never talk about, at least not to each other, this 360-degree crew had that in common.

●◑○○○○◐●

When Shepard's crew arrived at the Lunar Receiving Laboratory to wait out the rest of their quarantine, they brought along a haul of rocks and photographs that made the geologists ecstatic. But their enthusiasm was tempered by the fact that Shepard and Mitchell hadn't documented their finds as well as the scientists had expected. On the moon, Mitchell radioed the geologists that one of the rocks he collected would be easy to recognize because it had "a definite shape." But as one geologist told a reporter, all rocks have a definite shape, and in the LRL, Mitchell could not identify this one among the other samples. And although he and Shepard had trained to photograph rocks in place before collecting them, they had only documented a few samples in this manner. In the weeks to come, the scientists would have to try to locate the rest on the astronauts' panoramic photos. ☾

Meanwhile, using standard triangulation methods, some of the scientists had analyzed the pictures from Shepard and Mitchell's climb up Cone crater and had plotted the path of the two men. The results showed that Shepard and Mitchell had come within only 65 *feet* of the rim. Mitchell was furious when he heard. Ironically, he and Shepard had come the closest after they abandoned the search, when they walked down the slope to the strange

white rocks. If they had continued another 20 yards to the northwest,
they would have been staring into the enormous pit. Shepard and Mitchell had already paid up on the bet with Cernan and Engle, but the geologists did them one better: they sent a case of scotch into quarantine. As far
as they were concerned, 65 feet was close enough. Gordon Swann got a laugh
from Shepard when he said during a debriefing, "You weren't lost, and
you didn't know it!"

One morning around the same time, Shepard made his own discovery.
He was reading the paper at breakfast, when he noticed an article: "Astronaut
Does ESP Experiment on Moon Flight." Shepard laughed. "Hey, Ed," he
called out, "did you see this? This is the funniest goddamn thing I ever saw."

Mitchell came over to the breakfast table and looked at the article. "I did
it, boss," he said.

Shepard stared back at him in silence. He could barely believe Mitchell
had managed to get away with such a stunt without him finding out about it.
If he'd known about it before the flight, he would have hit the ceiling—*Not
on* this *mission, you're not*—but right now he was feeling pretty damn good.
He'd had his full-up mission. He could afford to be magnanimous. Why let a
little thing like this spoil a good breakfast?

●●○○○○●●

Not long after he got out of quarantine, Stu Roosa went to Downey to thank the workers at North American who had built the Apollo 14 command module. In a comment that rightfully belonged not only to the astronauts but to mission control, one manager told Roosa, "You saved the space program."

But within the scientific community, that news was greeted with skepticism. Ever since the first lunar landing, NASA had been criticized for failing to include scientists on Apollo missions. On the day Apollo 14 arrived in lunar orbit, a leading figure in lunar science, Caltech geologist Eugene Shoemaker, was in London to provide televised commentary on the mission for the BBC. At a press conference with an embittered Gordon Cooper, Shoemaker blasted NASA for doing a "completely miserable job" in integrating scientific goals into the moon program. To Shoemaker, who had always believed that the purpose of sending humans into space was to make discoveries, the potential of this $24 billion undertaking was being wasted. ☾

NASA was well aware of the criticisms voiced by Shoemaker and other scientists. Only three more lunar landings remained, but they had been planned to answer the scientists' frustrations. With those missions, Apollo would figuratively and literally reach its greatest heights.

Standing near the foot of *Antares*'s ladder, Shepard aimed his camera at the crescent earth, 244,000 miles away. From this distance the home planet looks four times larger than a full moon does from earth.

Late in Apollo 14's second moonwalk, Al Shepard adjusts the television camera set up to record activities near the lunar module *Antares,* resting on the Fra Mauro highlands. Beyond Shepard is the crater called Old Nameless.

Mitchell studies a photomap during the ascent of Cone crater. A folding sun shield, a feature used for the first time on Apollo 14, extends from his helmet, shading his eyes like the bill of a baseball cap.

Al Shepard makes the flag appear to flutter after planting it with Ed Mitchell during their first moonwalk at Fra Mauro on February 5. Shadows cast by the top of the lunar module *Antares* and two landing-leg support struts flank Shepard on the left; on the right is the shadow of the S-band antenna used to communicate with earth.

Twin tracks from the modular equipment transporter stretch across the Fra Mauro highlands toward *Antares*. Shepard took this photograph standing where he and Mitchell set up Apollo 14's package of scientific experiments.

CHAPTER 6: SAILORS ON THE OCEAN OF STORMS

I: The Education of Alan Bean

14 *Apollo had given the country the technology to go to other worlds:* Sources for this section: John Logsdon's unpublished manuscript on the transition from the Apollo to post-Apollo era (John Logsdon, Center for Space Policy, George Washington University, Washington, D.C.); Levine, *Managing NASA in the Apollo Era;* the author's interviews with former NASA administrators Tom Paine and James Beggs, and other NASA officials.

15 *In the LRL, five geologists:* The five were P. R. Bell and Elbert King of NASA's Lunar Receiving Laboratory, Ed Chao of the U.S. Geological Survey, Clifford Frondell of Harvard, and Robin Brett, formerly of USGS. See Cooper's *Moon Rocks* and Wilhelms's *To a Rocky Moon.*

21 *"Flight, try S-C-E to Aux":* As described on p. 375 of Murray and Cox, *Apollo: The Race to the Moon,* the Signal Condition Equipment was an electronics box that "performed an obscure role in translating the information from the sensors [onboard the spacecraft] into the signals that went to displays in the spacecraft and on the ground." When set to primary, the SCE would turn itself off under low-voltage conditions like those that occurred after the lightning strike. To turn it back on, the astronauts had to change the setting on the SCE to auxiliary. This resumed the flow of data to mission control.

23 *"Was that ever a sim they gave us!":* "Sim" was a commonly used abbreviation for simulation.

24 *"Look down there; those are campfires": Life,* December 19, 1969, p. 36.

25 *Conrad . . . took to calling him "Animal":* The nickname was taken from a character in the movie *Stalag 17,* about American soldiers in a German prisoner-of-war camp in World War II.

29 *For quite a while Bean felt like a minnow:* Looking back, Bean says, he shouldn't have blamed Shepard for his difficulties in the Astronaut Office. While Shepard did enjoy being ominous, he adds, it was unlikely that he was trying to harass Bean or any of the Fourteen. Bean says, "It's like the line about 'Don't tug on Superman's cape.' I was tugging." In particular, he was voicing opinions Shepard didn't agree with—and in retrospect, Bean says, those opinions were uninformed.

29 *the Apollo Applications Project, the space station planned for earth orbit in the 1970s:* The Apollo Applications Project was a key element in NASA's post-Apollo planning in the mid-1960s. As originally conceived by George Mueller, it would make use of Apollo hardware for an ambitious program of missions emphasizing space science and including space stations in earth orbit and long-duration visits to the lunar surface. One of Mueller's prime motivations was to preserve Apollo's 400,000-member government and industry team. But as NASA historian David Compton wrote, "Mueller faced a cruel paradox: the buildup of the Apollo industrial base left him no money to employ it effectively after the lunar landing." Before AAP could begin, its funding was slashed by Congress. The single surviving element, the space station, was later renamed Skylab. See Compton, *Living and Working in Space: A History of Skylab.*

31 *Electronics Test, the boondocks of test flight:* The reason Electronics Test was considered the boondocks, Bean says, is because it was where the majority of test pilot graduates were assigned, and it would have been easy to get lost there.

37 *a young mathematician named Emil Schiesser made a breakthrough:* Described on p. 383 of *Apollo: The Race to the Moon.*

II: Shore Leave

42 *dancing their way to the moon:* To make the picture complete, imagine the three men wearing matching caps, which Conrad had broken out after the Translunar Injection burn and presented to his crew. They were made of white Beta-cloth and they were personalized just like the Corvettes, with "CDR," "CMP," and "LMP." Conrad's had a little Teflon propeller on top.

45 *He looked down at what seemed to be a string of small volcanoes:* In reality, they were probably irregularly shaped impact craters, which can resemble volcanic features.

46 *"Where do you want me to put you?":* Conrad's conversation with Dave Reed is described on p. 385 of *Apollo: The Race to the Moon.*

51 *The gauge that was supposed to display lateral motion seemed to be broken:* Only when he was back on earth would Conrad find out that the display was working fine; he had put *Intrepid* in a near-perfect vertical descent.

54 *"that may have been a small one for Neil, but it's a long one for me":* According to Conrad, he was never able to collect his five hundred dollars.

58 *the first full-fledged scientific station to be set up on another world:* The first ALSEP was almost doomed before it could start transmitting data. The station's power generator was fueled by a cask of plutonium. When Bean tried to remove the cask from its graphite storage case on the side of the lunar module, it wouldn't budge. Only with help from Conrad—who beat on the case with his hammer, so hard that the case cracked—did Bean finally pull the cask free. Powered by its nuclear-electric generator, the ALSEP was designed to relay data for years.

64 *Gordon was jumping up and down: Life,* December 19, 1969, p. 37.

65 *"Now's the time to think up all sorts of fancy prose":* Ibid.

III: In the Belly of the Snowman

67 *The right leg of his space suit . . . was slightly too short:* It was Conrad's own fault, though that was no comfort. About a week before launch, technicians had found an air leak where the right boot joined the leg. The suit was quickly flown back to its manufacturer, the International Latex Corporation in Dover, Delaware, where it was repaired and returned to the Cape. Conrad was then called in to have the suit refitted. He wanted to wear his water-cooled underwear for the fitting, because it had thick plastic tubes that ran along the shoulders. But the water-cooled garment had long ago been packed in the LM. Couldn't he wear his training set? Conrad asked. Absolutely not was the answer; nothing that wasn't pristine could be worn inside a flight-ready suit. So Conrad did the fitting in a pair of flight-qualified plain cotton long johns. He tried to leave enough room for the bulk of the plastic tubes, but he underestimated. He was only off by a quarter of an inch, but that was enough to make a difference.

73 *Each step launched him into the air for long seconds:* Bean notes that while it seemed as if he were going 10 feet with each step, in reality his strides were probably the same length as when he ran on the beach at the Cape. He says that in one-sixth g, his body was so light that he didn't have much traction, and that prevented him from pushing off very hard.

74 *Conrad, running ahead of him, managed to study the map even as he bounded along:* Despite Conrad's mastery of the "lunar lope," he became the first astronaut to fall down on the moon, while collecting samples during this moonwalk. He got up easily by grabbing Bean's hand. Most of the time, however, Conrad and Bean found they could avoid falling because in one-sixth g it happened so slowly that they could, by running, get their feet under themselves and stand up.

75 *As they ran . . . they were beginning to feel the strain of their adventure:* Some of the exertion was due to the stiffness of the waist joint of their suits, a problem that was remedied on the last three Apollo missions.

78 *While Conrad held the tool carrier, Bean rummaged among the samples:* Conrad and Bean don't agree about who did what. This version was reconstructed with help from Eric Jones of Los Alamos National Laboratory, who went over the caper with Conrad and Bean while researching his forthcoming Lunar Surface Journals.

85 *Where Bean's cold had come from, he had no idea:* Later it would turn out that both men had stuffy heads from zero gravity, which causes the fluids to migrate to the upper body.

87 *the sights had been the most spectacular of his life:* There were spectacular sights even on the relatively boring trip back to earth. On the last day of the mission the men became the first to witness an eclipse of the sun by the earth. For over an hour, as *Yankee Clipper* flew through the earth's shadow, they saw their home planet ringed by a rainbow of sunlight shining through the atmosphere. In the middle of the darkened world they could see a faint glow; it was light from the full moon, shining on a midnight ocean.

CHAPTER 7: THE CROWN OF AN ASTRONAUT'S CAREER

I: A Change of Fortune

95 *two more Apollo flights were in jeopardy:* At this time, NASA was still planning to fly Apollo 18 and 19 but had postponed them until 1974, after the missions to the Apollo Applications (later Skylab) space station. See Compton, *Living and Working in Space: A History of Skylab,* p. 135.

95 *Bob Gilruth had privately called for an end to the moon landing program:* A number of people remember Gilruth expressing this sentiment. One of them is Chris Kraft. In 1989 Kraft recalled, "He said to me on several occasions [after the first lunar landing], 'We've done that. It's too risky. We're liable to lose one of these things, and we just can't afford to have that happen. . . . Why should we go anymore? We've got the rocks.' " Kraft added, "I personally didn't feel as strongly as Gilruth did about not flying again." Kraft believed Gilruth felt the way he did in part because he was near the end of his career, and age tends to temper the willingness to take risks. "He may even be right," Kraft said. "Experience is a great teacher."

114 *To convert the data from* Odyssey's *platform to* Aquarius's *frame of reference:* The work was complicated by a drastic change in reference. To visualize this, remember that the nose of the command module was docked to the lunar module's ceiling. An astronaut seated in the command module was oriented 90 degrees to his crewmates standing in the LM cabin. As a result, the three degrees of rotation possible for the joined spacecraft—roll, pitch, and yaw—looked different, depending on which of the two craft he was in. If *Odyssey* pitched up or down, then by Lovell and Haise's reckoning, so did *Aquarius.* But if *Odyssey* spun along its axis—which Swigert called roll—Lovell and Haise felt *Aquarius* yaw, that is, nod to one side. And when *Odyssey* wagged from side to side—a yaw maneuver—it looked like roll to Lovell and Haise.

II: The Moon Is a Harsh Mistress

119 *Haise and a navy pilot named Edgar Mitchell:* Ed Mitchell remembers that he and Haise became "great friends and great rivals" during their time at Grumman. They would take turns quizzing each other about esoteric details of the LM's design and operation. And they played one-upmanship any time one of them got a step closer to flying in space. When Mitchell was named to Apollo 13, he appeared to have won the competition, but when the Apollo 13 and 14 crews were swapped, for reasons described in Chapter 8, Haise got the victory. Ironically, the victory placed Haise in a struggle for his life, while Mitchell spent hour after hour in the lunar module simulator, working out procedures for his friend to use aboard *Aquarius.*

119 *to perform in ways it had not been designed for:* The LM lifeboat idea dated back to 1961, before the preliminary design for the lander had even been drawn up. Grumman engineers had envisioned using the lander's engines to push a crippled command ship out of lunar orbit. Eight years later, training for Apollo 9, Gene Kranz's flight controllers were hit with a simulation in which Jim McDivitt and Rusty Schweickart were stranded in a

lopsided orbit around the earth. They had to devise ways to stretch the LM's supplies to last 30 hours instead of 18, long enough for Dave Scott to rescue them. While that exercise foreshadowed the effort to save Lovell's crew, no one ever envisioned using the LM in precisely the way it was used on Apollo 13 (sources: the author's interviews with Gene Kranz; Murray and Cox, *Apollo: The Race to the Moon*, p. 423).

120 Power values for radio and environmental control system: These numbers were supplied by former flight controller Robert Legler. Specifically, 1.29 amps is for the S-band transceiver; 5.5 amps is the power needed to run one suit fan.

122 *Mattingly blushed and jingled the change in his pocket: Newsweek*, April 13, 1970, p. 63.

125 *Lovell could only imagine what Swigert was thinking now:* Apollo 13 was halfway to the moon before Swigert realized he had not filed his income taxes and that he would be quite unable to do so before the April 15 deadline. The subject came up as scientist-astronaut Joe Kerwin was reading the Sunday morning news: "Today's favorite pastime across the nation—Uh oh, have you guys completed your income tax?"

Swigert radioed, "How do I apply for an extension?" Mission control exploded with laughter. "It ain't too funny, things happened real fast down there and I do need an extension. I'm really serious . . ."

"You're breaking up the room down here," Kerwin said. A few minutes later he assured Swigert that there wouldn't be any problem: an automatic extension is granted to anyone who is out of the country at tax time.

127 *She'd asked some of the NASA people about the odds of saving the men:* For his own part, Chris Kraft says he would never have given Marilyn Lovell such a dismal forecast. He says that once the LM lifeboat plan was adopted he never had any doubts of getting Lovell's crew home. "It's not very romantic," Kraft says, "but it's true!"

130 *worse than they were willing to admit:* These doubts are based on some notes taken by Chris Kraft at one of the astronauts' post-flight debriefings. The notes do not specifically identify the astronaut who made the comment. Today, Lovell (along with Haise) denies having had such thoughts, but it is possible that his feelings during the flight were different from those he had long afterward. And as Ken Mattingly points out, the flow of information from earth to Apollo 13 was imperfect enough to allow room for such thoughts—which he says would have been entirely natural—in the minds of Lovell and his crew.

III: The Chill of Space

153 *immediate problem of where to store urine:* If it became necessary—which it never did—Haise had even figured out a way to put urine to good use by transferring it into the LM's cooling system.

154 *another bit of ingenuity from Houston:* Another procedure devised in Houston by controller Robert Legler was especially critical: it allowed Lovell's crew to charge *Odyssey*'s batteries using *Aquarius*'s power.

158 *After three days of constant background noise from Aquarius's fans and pumps:* Fred Haise, wanting to preserve those sounds for posterity, recorded them with a hand-held tape recorder before *Aquarius* was jettisoned.

158 *A serious crack was another matter:* In Houston, some of the flight controllers had privately pondered the same thing and, like Haise, realized there was nothing they could do about it. See *Apollo: The Race to the Moon*, p. 443.

162 *As presidential adviser John Ehrlichman told historian John Logsdon years later:* John Logsdon's unpublished manuscript on the transition from the Apollo to post-Apollo era (John Logsdon, Center for Space Policy, George Washington University, Washington, D.C.).

162 *After Apollo 13 some of Nixon's advisers:* Nixon told this to Stu Roosa when the Apollo 14 astronauts visited the White House in 1971.

CHAPTER 8: THE STORY OF A FULL-UP MISSION

I: Big Al Flies Again

173 *"Why don't you fix your little problem and light this candle?":* Life, May 12, 1961.

175 *John Glenn was so angry:* The author's conversation with writer Howard Benedict, 1992.

176 *In April, representatives of the President's Advisory Committee:* The space subcommittee of PSAC, which Shepard recalls as being composed primarily of medical specialists, was chaired by MIT scientist Jerome Wiesner, who had been against the manned space program from the beginning. Shepard interviews with author; Shepard interview with the JFK Library Oral History Project, June 12, 1964 (JFK Library, Boston); Murray and Cox, *Apollo: The Race to the Moon*, pp. 66-67, 70-71.

180 *Shepard was on the way to becoming a millionaire:* Prior to Apollo 14 there were occasional rumors stating that Shepard was already a millionaire, but he denied them.

184 *"If you guys don't mind flying with an old retread":* It was Pete Conrad who bestowed a new nickname on Shepard; it happened after the splashdown of Apollo 12, when Conrad, Gordon, and Bean arrived at Ellington Air Force Base in their quarantine trailer. Behind the glass, Conrad looked out at the well-wishers who had come to meet them, and now he saw the un-grounded Al Shepard. He grinned his gap-toothed grin, grabbed the PA microphone, and announced, "Here comes the rookie!" Shepard flashed Conrad a look that could have melted lead. Conrad was glad he'd said it. And it wasn't long before some began referring to Shepard, Roosa, and Mitchell as the "all-rookie crew."

185 *And as far as Deke Slayton was concerned, if Shepard wasn't qualified, no one was:* Slayton's confidence in Shepard is illustrated by the fact that Apollo 13 would have been NASA's last chance to make a lunar landing in 1969 if 11 and 12 had failed.

186 *"They ought to hire tiddlywinks players as astronauts":* Newsday, February 5, 1971.

186 *the only one of the Original 7 to reach the moon:* This was contrary to the expectations of many observers, who had initially thought that one of the Original 7 would probably become the first man on the moon. In the mid-1960s, some of the geologists who trained the astronauts made four predictions of who would most likely command the first landing. Three of their picks were members of the Original 7: Wally Schirra, Scott Carpenter, and Gordon Cooper. (The fourth, and the only one who made it to the moon, was Pete Conrad.)

186 *Now he had been saved . . . by George Mueller:* By this time, however, Mueller was no longer at NASA, having left the agency at the end of 1969 (Levine, *Managing NASA in the Apollo Era*, p. 308).

187 *At the end of August . . . canceled two more Apollo missions:* In a 1989 interview, Tom Paine, who had participated in the decision to cancel these missions, said that he had been faced with the dilemma of whether to continue exploring the moon at the expense of NASA's long-term future. True, the scientists were urging him not to trim any more missions, but to Paine Apollo seemed to be nearing a point of diminishing returns. Was it worth deferring future projects to land on the moon a few more times? Paine decided it was not. Under the circumstances, Paine felt, he had made the best of things.

It is clear that in addition to the budgetary and scientific issues, the risks involved in the lunar missions—to human lives and to the future of NASA—was an important factor in the decision, but it is difficult to know how much of a factor. In 1989 Chris Kraft said that he and his colleagues "put all that in a pot and stirred it up. And the fathers of NASA and leaders of Congress concluded: Stop after [Apollo]17 instead of [Apollo] 19."

For the scientists, the cancellations were a lesson in lost opportunities. Not until September 1970, after the decision had been made, did they voice protest. In the middle of that month thirty-nine lunar scientists wrote a letter to Congressman George P. Miller, the chairman of the House Committee on Science and Astronautics and a long-time supporter of the space program. Canceling the final three Apollo flights, they wrote, might cause the lunar science program to "fail in its chief purpose of reaching a new level of understanding" about the moon and about our own planet. Miller replied by stating that he and his committee had tried to get an additional $220 million for Apollo into the authorization bills for 1970 and 1971, but that "the Nixon administration, in realigning national priorities, has relegated the space program to a lesser role." However, he told the scientists, "Had your views on the Apollo program been as forcefully expressed to NASA and the Congress a year or more ago, this situation might have been prevented." See Compton, *Where No Man Has Gone Before: A History of the Apollo Lunar Exploration Missions,* p. 203. Other sources on the end of Apollo: Compton, *Living and Working in Space: A History of Skylab;* Levine, *Managing NASA in the Apollo Era;* John Logsdon's unpublished manuscript on the transition from the Apollo to post-Apollo era. See also the author's article "Why Haven't We Been Back," *Air & Space/Smithsonian,* July 1989, pp. 90-97.

188 *Their husbands were playing around:* To be sure, a number of astronaut wives already knew of, or suspected, their husbands' extramarital affairs. One remarked in 1967 that she accepted a certain amount of infidelity the way she thought of speeding tickets: "It's bound to happen once in a while." That the astronauts' indiscretions weren't covered by the media is a reflection of their special place in 1960s America. As *Life*'s Dora Jane Hamblin wrote in 1977, "I think *Life* treated the men and their families with kid gloves. So did most of the rest of the press. These guys were heroes. . . . I knew, of course, about some very shaky marriages, some womanizing, some drinking, and never reported it. The guys wouldn't have let me, and neither would NASA" (unpublished article by former NASA public affairs officer Paul Haney; Hamblin letter to Perry Michael Whye, Iowa State University, January 1977).

188 *the modifications to the command and service modules:* To safeguard against a repeat of the Apollo 13 accident, engineers redesigned the service module's oxygen tanks and increased their number from two to three. They also installed valves between tanks to prevent any single rupture from depleting more than one tank. In the command module, they added a contingency water storage system so that astronauts would have drinking water if the command module were disabled. In the lunar module, they added an extra battery for emergency power. *Apollo Program Summary Report,* 1975, NASA Lyndon B. Johnson Space Center (internal publication).

189 *Shepard sketched a design that showed an astronaut pin:* The astronaut insignia showed a three-tailed star ascending through a ring.

189 *when angered he was capable of outbursts of temper:* Once during a training exercise, Mitchell became frustrated with a balky experiment and shook it so hard that it broke (Associated Press news story, January 31, 1971, published in the *Washington Star*).

189 *Mitchell . . . seemed to be carrying the load for Shepard with the lander's systems:* Mitchell says Shepard leaned on him a great deal during training. By agreement with Shepard, Mitchell played the role of instructor pilot with his commander. At first, Mitchell coached Shepard through malfunction procedures; then, as Shepard caught on, Mitchell hung back. In the simulator one day a few months before launch, Mitchell turned to Shepard and said, "Okay, Boss. I believe you're ready to go."

193 *at age forty-seven:* At the age of forty-seven years and two months, Shepard was the oldest American to go into space up to that time. Soviet cosmonaut Georgiy Beregovoy was four months older when he made his Soyuz 9 flight in 1968 (Hawthorne, *Men and Women of Space*).

193 *Okay, buster, you volunteered for this thing:* Carpenter et al., *We Seven,* p. 195.

193 *a booster a hundred times more powerful:* Shepard's Redstone booster had 78,000 pounds of thrust; the first stage of the Saturn V had more than 7.5 million pounds of thrust.

II: To the Promised Land

194 *The culprit might be something as simple as a tiny piece of debris on the mechanism:* The cause of the problem was never determined. In their studies following the flight, engineers ruled out an ice particle as the culprit, but did list contamination or debris as possibilities.

197 *Mitchell was probably the only astronaut who missed the presence of psychologists:* MacKinnon and Baldanza, *Footprints*, pp. 91-92.

199 Mitchell's ESP experiment: The results of Mitchell's experiment do not bear easy interpretation. Because Apollo 14's launch happened forty minutes behind schedule, so did Mitchell's sleep periods on the way to and from the moon, throwing off the timing of his experiment. On earth, the test subjects were actually trying to "receive" forty minutes before Mitchell was trying to "send." Despite this fact, Mitchell says, the experiment produced useful results. When the subjects tried to imagine the picture that Mitchell was thinking of at any given moment, they were often wrong in their guesses—but they were wrong, Mitchell says, significantly more often than pure chance would have dictated. Mitchell says this suggests that their subconscious minds knew something was wrong (that is, the experiment wasn't happening according to plan) and, in a precognitive way, adjusted to the situation by giving wrong answers. The missed guesses, Mitchell says, led the press to call the experiment a failure. But to Mitchell, it demonstrated that psychic phenomena were indeed at work.

199 Beginning at 2 days, 6 hours, and 57 minutes into the flight, the Mission Elapsed Time clock was advanced 40 minutes, to make up for the 40-minute delay in Apollo 14's launch. Mission Elapsed Times of events in Chapter 8 reflect this change.

201 *Once more, with help from mission control, they had made a narrow escape:* As evidence of his and Shepard's determination to make the landing, Mitchell recalls that in simulations, even when the instructors made the LM's computer fail entirely, they had still landed. Mitchell would handle the yaw and roll while Shepard controlled pitch and rode the throttle. The simulator didn't "fly" very well that way—they made some awfully hairy approaches—but they got it down in one piece.

IV: The Climb

212 *scientists were divided into two main camps:* For a definitive account of the history of lunar science, see Wilhelms, *To a Rocky Moon: A Geologist's History of Lunar Exploration.*

216 *firmly perched on the rolling hills of Fra Mauro:* Just to be sure, Shepard and Mitchell rigged a plumb line from a piece of string tied to a handhold, to use as an attitude reference; the 8-ball and computer were turned off. *Antares* wasn't tipping over; the sound that awoke them proved to be a noisy valve.

216 *a mocking version of the Apollo 14 mission patch:* On the patch, the rendering of the Coyote is a satirical reference to Shepard's crew. For red-headed Stu Roosa, the Coyote has red fur. For Ed Mitchell, who was not big on exercise, he has a pot belly. And he has a long gray beard, for Old Man Shepard.

220 *With or without the MET, they would get there:* One other reason Mitchell wanted to reach the crater itself: he wanted to roll a rock into it. On geology field trips, rock-rolling was a hallowed ritual which, in its purest expression, saw two or three pilots lying on their backs, pushing against a massive boulder with their boots until it went crashing down a hillside among the cedar trees, or into the steaming throat of a volcano.

224 *The deepest rocks, and perhaps the most important, would lie near the crater's edge:* Cone crater lay atop a ridge that was, like the rest of the Fra Mauro hills, believed to consist of debris ejected from a giant impact crater called the Imbrium basin, 340 miles to the north. Finding the date of Imbrium's formation, and sampling the rocks it tore out of the lunar crust, was one of the geologists' top priorities. The rocks near Cone's rim were most likely to be true samples of Imbrium ejecta, and therefore, potentially the most important.

224 *"Oh, let's give it a whirl!":* The tone in Mitchell's voice was uncharacteristically unrestrained, and frustration wasn't the only reason. After Conrad and Bean broadcast lively chatter during their moonwalks, Mitchell's family and friends, anticipating his own flight, teased him for being too serious. He ought to follow Conrad's and Bean's example, they said. Mitchell says that he had that in the back of his mind as he and Shepard were climbing Cone crater: he was playing to the gallery just a bit.

229 *Ed Mitchell felt a weariness:* Mitchell's emotional state aside, he and his crewmates were physically spent. After Apollo 13, mission planners weren't about to let Shepard's crew linger in lunar orbit any longer than necessary, and as soon as Shepard and Mitchell had rejoined Roosa the three men had pushed through the procedures for jettisoning the LM, and then, the Transearth Injection. By the time they bedded down for the night it had been a twenty-two-hour day.

232 *a haul of rocks and photographs that made the geologists ecstatic:* Stu Roosa was told that his pictures of Descartes were so good that they actually exceeded the theoretical resolution of the lens.

232 *Shepard and Mitchell hadn't documented their finds as well as the scientists had expected:* "Letter from the Space Center," by Henry S. F. Cooper, *The New Yorker,* April 11, 1971, pp. 126-27.

234 *Shoemaker blasted NASA for doing a "completely miserable job":* Newsday, February 5, 1971.

ACKNOWLEDGMENTS

The editors of this book wish to thank the following persons and institutions for their assistance.

Gay Alford, Houston, Tex.
David Burgevin, Smithsonian Institution, Washington, D.C.
Paul Calle, Stamford, Conn.
Bob Craddock and Tom Crouch, Smithsonian Institution, National Air and Space Museum, Washington, D.C.
Maureen M. Dilg, National Geographic Society, Washington, D.C.
Jan Evans, Scottsdale, Ariz.
Christopher J. Faranetta, Energia Ltd., Alexandria, Va.
Mike Gentry and Kathy Strawn, NASA/Media Services, Houston, Tex.
Bob Green, The Greenwich Workshop, Shelton, Conn.
George Hendry and Larry Felieu, NorthropGrumman History Center, Bethpage, N.Y.
Timothy Hughes and Marc Pompeo, Rare & Early Newspapers, Williamsport, Pa.
Sherie Jefferson, Irene Jenkins, and David Sharron, Information Dynamics Incorporated, Houston, Tex.

Melissa Keiser, Smithsonian Institution, National Air and Space Museum, Washington, D.C.
The Lovell Family, Lake Forest, Ill.
Eric Newton & Jeffrey Schlosberg, Newseum, Arlington, Va.
Margaret Persinger, Kennedy Space Center, Fla.
Gwen Pittman, NASA Headquarters, Washington, D.C.
Gary Pressel, Adtech Photo Imaging, Houston, Tex.
Christopher Roosa, Falls Church, Va.
Jack Roosa, San Antonio, Tex.
Joan Roosa, Gulfport, Miss.
Rosemary Roosa, Biloxi, Miss.
Kipp Teague, Project Apollo Archive, Lynchburg, Va.
Mary Weeks, Lake Forest, Ill.
Beth Williams, Seabrook, Tex.

PICTURE CREDITS

The sources for the illustrations in this book appear below. Credits from left to right are separated by semicolons, from top to bottom by dashes.

Spine: NASA—NASA #S6750531. **Cover:** NASA #AS08142383; plaque created by John Drummond © Time-Life Books, Inc.; NASA #AS12466716 (inset).

1: NASA #AS08142383; plaque created by John Drummond © Time-Life Books, Inc.; NASA #AS12466716 (inset). **2, 3:** Plaque created by John Drummond © Time-Life Books, Inc.; NASA #AS08142383. **4:** Marvin Chaikin. **6, 7:** NASA #12476891. **8, 9:** NASA #AS12466780. **10, 11:** NASA #AS12487149. **12:** NASA #AS12497278. **13, 95, 173:** Detail from NASA #S6818733 **14:** C. Bonestell, courtesy Andrew Chaikin. **15:** Walter Sanders. **16:** A. Patnesky/National Geographic Society Image Collection. **18:** NASA #S6959430. **22:** NASA. **23:** Lee Balterman/*Life* Magazine © Time Inc. **25:** Leonard McCombe/*Life* Magazine © Time Inc. **26:** NASA #S6654455. **27:** Courtesy Pete Conrad. **28:** Courtesy Alan Bean. **30:** Courtesy Beth Williams. **34, 35:** Courtesy Alan Bean. **36:** Movie Stills Archives, Harrison, Nebr. **38:** Courtesy Rocket Space Corporation Energia Korolev Russia. **39:** Smithsonian Institution, Neg. No. 97-15886/2; NASA #AST0296—NASA, Smithsonian Institution, courtesy RSC Energia. **40:** NASA. **43:** NASA, courtesy Andrew Chaikin. **44-45:** NASA, courtesy Andrew Chaikin. **48, 49:** NASA #AS12517507. **50:** NASA, courtesy Andrew Chaikin. **52, 53:** NASA #AS12466716. **54:** NASA #AS12476949. **55:** NASA #AS12466729, courtesy Andrew Chaikin. **56:** Courtesy Pete Conrad. **57:** NASA #AS12466795. **58:** NASA #AS12466790. **60, 61:** NASA #AS12466806. **62:** NASA #12476919. **64:** NASA #S6956702. **66:** NASA. **68:** CORBIS/UPI; Ralph Morse/*Life* Magazine © Time Inc.; Fritz Goro/*Life* Magazine © Time Inc. **69:** Art by Paul Calle © 1969 Time-Life Books, Inc. **71:** NASA #S6959538. **72-73:** NASA #AS12497209; #AS12497210; #AS12497211; #AS12497212; #AS12497213; #AS12497214; #AS12497215. **74:** NASA #AS12497286. **77:** NASA #AS12497318, courtesy Kipp Teague, Project Apollo Archive. **80, 81:** NASA #AS12487133. **83:** NASA #S8037406. **84:** Lee Balterman/*Life* Magazine © Time Inc. **86:** NASA #S6960354. **88-93:** Alan Bean. **94:** Ralph Morse/*Life* Magazine © Time Inc. **96:** NASA #7035140. **98, 99:** NASA #7035139. **101:** NASA #7036485. **103:** NASA #S6960662. **106:** NASA, courtesy Andrew Chaikin. **108, 109:** Alan Bean, courtesy Greenwich Workshop. **110:** John P. Filo. **112:** Art by Time Life Inc. based on NASA diagram. **115:** NASA

#S7035014. **118:** NASA #7034412. **120:** NASA #7034902. **123:** Everett Collection, New York. **124, 125:** NASA #7034847. **126:** NASA, courtesy Andrew Chaikin. **127:** Bill Eppridge/*Life* Magazine © Time Inc.—Ralph Morse/*Life* Magazine © Time Inc. **128, 129:** Bill Eppridge/*Life* Magazine © Time Inc. **132, 133:** NASA #AS13618727. **134:** NASA, courtesy Andrew Chaikin. **136:** Courtesy Ruth Goldberg. **138-139:** NASA, courtesy Andrew Chaikin. **142, 143:** NASA #S7035013. **144:** NASA #AS13628929. **146-147:** NASA, courtesy Andrew Chaikin. **150-151:** NASA, courtesy Andrew Chaikin. **151:** NASA, courtesy Andrew Chaikin. **152:** NASA #AS13608588. **155:** Walter Iooss/*Sports Illustrated*. **156:** NASA #AS13598500. **160:** Bill Eppridge/*Life* Magazine © Time Inc. **161:** Ralph Morse/*Life* Magazine © Time Inc.—CORBIS/UPI. **163:** NASA #AS13598513. **164, 165:** NASA #7035652. **166, 167:** NASA #S7015819. **168, 169:** NASA #S7035614. **170, 171:** Ralph Morse/*Life* Magazine © Time Inc. **172:** NASA #S7045697. **174:** Ralph Morse/*Life* Magazine © Time Inc. **177:** Ralph Morse/*Life* Magazine © Time Inc. **179:** CORBIS/UPI. **180:** NASA #S6650713. **182:** Courtesy Jack Roosa. **184:** Courtesy Gay Alford. **185:** NASA #S6546366. **187:** Ralph Morse/*Life* Magazine © Time Inc. **189:** NASA #S7017851. **190, 191:** NASA #AS14669232, courtesy Kipp Teague, Project Apollo Archive. **192:** NASA #S7117620. **195:** © George Butler, New York. **196:** NASA #S7117141—NASA #S6850869. **198:** NASA; Arnold Zann; NASA #AS14729937 (background). Courtesy of the Rhine Research Center, Durham, N.C. (symbols). **201:** NASA, courtesy Andrew Chaikin (2). **202:** NASA, courtesy Andrew Chaikin. **204:** NASA #AS14669230. **206-207:** NASA #AS14669305; #AS14669304; #AS14669306. **209:** NASA #AS14669344. **210:** Courtesy Rosemary Roosa. **211:** NASA, courtesy Andrew Chaikin. **212:** Michael Ochs Archives, Venice, Calif. **214:** Don Davis. **217:** Courtesy Jan Evans. **218, 219:** NASA #AS14689405. **221:** NASA, courtesy Andrew Chaikin. **222, 223:** NASA/National Geographic Society Image Collection. **224:** NASA, courtesy Andrew Chaikin. **226, 227:** NASA/National Geographic Society Image Collection. **228:** NASA #AS14689452. **230, 231:** Alan Bean. **233:** NASA #S7120110. **235:** NASA #AS14649197. **236, 237:** NASA #AS14689486. **238, 239:** NASA #AS14649088; #AS14649089. **240, 241:** NASA #AS14669232, courtesy Kipp Teague, Project Apollo Archive. **242, 243:** NASA #AS14679367.

INDEX

Shepard, Alan, 172-181, *177, 180,* 183-195, *187,* 199-203, 216-233, *233;* business ventures, 179-180; emblem designed by, *189;* Gemini project, 178-179; illness of, 179, 180-181, 184; Mercury project, 173-175, *174;* moodiness of, 29, 183-*184,* 188-189; on the moon, *204, 218-219, 222-223, 230-231, 236-237, 240-241;* speech on moon landing, 202; training, *172*

Shoemaker, Eugene, 234

Simulators, Apollo 12 training, 36, *40, 64;* Apollo 13 training, 100-101, *118;* Gemini simulators, *30;* lunar module simulators, *187, 201*

Sjoberg, Sig, *142-143*

Slayton, Deke, 101, *120, 142-143, 177,* 185-186

Sleep in space, 27, 65-*66,* 67, 137-138, 140, 145-*146,* 149, 150-151

Snowman, 44, *71*

Soil. *See* Lunar soil

Soviet Union, space program, *38-39;* U.S. attack plans for, 181-182

Space, navigating in, 23-24, 130-131; shaving in, *44;* sleeping in, 27, 65-66, *66,* 67, 137-138, 140, 145-*146,* 149, 150-151

Spacecraft, Soviet, *39. See also* Equipment malfunctions; Lunar modules; Rockets; *individual spacecraft by name*

Space docking, 194-195, *196. See also* Space rendezvous

Space programs, Soviet, *38-39. See also* National Aeronautics and Space Administration (NASA)

Space race, *38-39*

Space rendezvous, 64-65, 82-85. *See also* Space docking

Space station project, 13-14

Space suits, commander's suits, *205;* construction of, *69;* discomfort of, 65, 67; EMU checks, 75-76; history of, *68-69;* pressure increases in, 76; Soviet, *39;* sun shields, *239*

Space Task Group (STG), 14

Spacewalks, 26-27, *26. See also* Moonwalks

Steinem, Gloria, *179*

Student demonstrations, *25, 110,* 187

Substitution of crew on Apollo 13, 100-101, 122-125

Sun, earth eclipsing, *83*

Sun shields for space suits, *239*

Surveyor 3 probe, *80-81;* as a landing target, *40,* 43-44, 46-47, *54,* 64; photo pose by, 76-79; television camera, 79, *80-81*

Swann, Gordon, 221-225

Swigert, Jack, *101,* 103-110, 111, *124-125,* 127-141, 145-154, 157-162, *166-169;* in flight, *126, 134, 147, 150*

T

Television camera from Surveyor probe, 79, *80-81*

Television transmissions, Apollo 13 landing, 159-161, *160-161;* about Apollo 14, 234; Apollo 14, 227-229, *236-237;* from the moon, *8-9,* 57, 67, 227-229, *236-237;* from space, *96-97, 98-99,* 102

Timelines, 25, 36, 110, 123, 155, 179, 195, 212

Tindall, Bill, *142-143*

Tool carriers for the moon, *77, 218-219,* 220, *222-223,* 225

Topeka Daily Capital Apollo 13 story, *136*

Tracks left by modular equipment transporter, *242-243*

Training for astronauts, delegation by commanders, 188-189; Gemini project, *30;* moonwalk training, *40, 172;* simulators, *30,* 36, *40, 64,* 100-101, *118*

U

Urey, Harold, lunar geology theories, 212

Urine disposal on Apollo 13, 153-154

U.S. Air Force aviation programs, 181-*182*

U.S. Navy aviation programs, 27-29, 31

U.S.S.R. space program, *38-39;* U.S. attack plans for, 181-182

V

Vietnam War protests, *25, 110,* 187, *195*

Von Braun, Wernher, *15;* Mars expedition plans, 14

W

Water supply on the *Aquarius,* 121, 146, 153

Wiley Post pressure suit, *68*

Williams, C. C., *30,* 31, 32

Windler, Milt, *142-143*

Winter, Lumen (works by), *103*

X

X-15 pressure suit, *68*

Y

Yankee Clipper, 43-45, 63-65, 82-85

Young, John, *120*

ANDREW CHAIKIN, A MAN ON THE MOON
VOLUME II: THE ODYSSEY CONTINUES

First published in 1994 by Viking Penguin, a division of Penguin Books U.S.A. Inc., as Book 2 of A Man on the Moon: The Voyages of the Apollo Astronauts by Andrew Chaikin.

This Time-Life edition is published by arrangement with Viking Penguin, a division of Penguin Putnam Inc.

Cover, text design, and captions © 1999 Time Life Inc.

Second printing 1999. Printed in U.S.A.

School and library distribution by Time-Life Education, P.O. Box 85026, Richmond, Virginia 23285-5026

TIME-LIFE is a trademark of Time Warner Inc. and affiliated companies.

Library of Congress Cataloging-in-Publication Data
Chaikin, Andrew, 1956-
A man on the moon / by Andrew Chaikin and the editors of Time-Life Books.
p. cm.
Includes index.
Contents: 1. One giant leap — 2. The odyssey continues — 3. Lunar explorers.
ISBN 0-7835-5675-6 (v. 1).—ISBN 0-7835-5676-4 (v. 2). —ISBN 0-7835-5677-2 (v. 3).
1. Project Apollo (U.S.)—History. 2. Space flight to the moon—History. I. Title.
TL789.8.U6A5244 1999
629.45'4'0973—dc21   99-15449
                                                      CIP

Time-Life Books is a division of Time Life Inc.

TIME LIFE INC.
PRESIDENT and CEO: George Artandi

TIME-LIFE BOOKS
PUBLISHER/MANAGING EDITOR: Neil Kagan
SENIOR VICE PRESIDENT, MARKETING: Joseph A. Kuna
VICE PRESIDENT, NEW PRODUCT DEVELOPMENT: Amy Golden

EDITOR: Lee Hassig
DIRECTORS, NEW PRODUCT DEVELOPMENT: Mary Ann Donaghy, Elizabeth D. Ward, Paula York-Soderlund

Design Director: Cynthia T. Richardson
Assistant Art Director: Janet Dell Russell Johnson
Senior Marketing Manager: Paul Fontaine
Project Manager: Karen Ingebretsen
Associate Editor/Research and Writing: Ruth Goldberg
Senior Copyeditor: Judith Klein
Page Makeup Coordinator: Kimberly A. Grandcolas
Editorial Associate: Patricia D. Whiteford

Special Contributors: Andrew Chaikin (captions); Marilyn Murphy Terrell (research); Christine Stephenson (text); John Drummond, Mary Gasperetti (design and production); Marianne Dyson, Amanda Stowe (picture research); Antheus L. Bowden (picture coordination); Sunday Oliver (index).

Correspondents: Maria Vincenza Aloisi (Paris), Christine Hinze (London), Christina Lieberman (New York).

Director of Finance: Christopher Hearing
Director of Book Production: Patricia Pascale
Director of Imaging: Marjann Caldwell
Director of Publishing Technology: Betsi McGrath
Director of Photography and Research: John Conrad Weiser
Director of Editorial Administration: Barbara Levitt
Manager, Technical Services: Anne Topp
Page Makeup Manager: Debby Tait
Senior Production Manager: Ken Sabol
Production Manager: Virginia Reardon
Quality Assurance Manager: James King
Chief Librarian: Louise Forstall

Separations by the Time-Life Imaging Department